Electromagnetic Field Theory

AN INTRODUCTION FOR ELECTRICAL ENGINEERS

R. D. STUART

Department of Electrical Engineering, Northeastern University

ADDISON-WESLEY PUBLISHING COMPANY, INC.

READING, MASSACHUSETTS · PALO ALTO · LONDON

NEW YORK · DALLAS · ATLANTA

This book is in the
ADDISON-WESLEY SERIES IN
ELECTRICAL ENGINEERING

Consulting Editors
DAVID K. CHENG AND WILLIAM A. LYNCH

Printed in the United States of America

Library of Congress Catalog Card No. 65-16888

Preface

This text is intended for use by electrical-engineering students in the second half of their undergraduate curriculum. It is assumed that they will have some familiarity with the experimental side of the subject from their physics courses in electricity and magnetism, and that they will at least have had a course in vector algebra. It is also assumed that they will take the courses in circuit theory before those in field theory. At present, the logical sequence for engineering students seems to be: lumped-parameter circuits, distributed-parameter circuits, and fields. The subject of transmission lines is accordingly not touched on here, but a knowledge of distributed-parameter circuits is not a prerequisite for the use of this book.

The first chapter is intended to get across the idea of a field. The discussion is in terms of electric currents. The use of currents and voltages in this chapter without definition is bound to upset the purist and those acclimatized to the conventional approach. However, as these were the everyday tools of the student in his circuit-theory courses, he will not share their concern. The extension of circuit theory to the "three-dimensional distributed circuit" is used to develop most of the vector calculus subsequently needed. Most engineering students find the operations of vector calculus a little difficult to grasp when taught from the point of view of pure mathematics. The use of electrostatics is no improvement, since the notions are equally abstract. Probably the most pictorial presentation can be given in terms of hydrodynamics. However, electrical-engineering students are not so familiar with this subject as they are with electric circuits, and furthermore they are not usually favorably impressed by discovering that before they can get on with their electrical engineering they must first study hydrodynamics. Hence we have chosen electric-current fields for this purpose.

The second chapter defines the electric quantities in terms of the fundamental quantities. The student is encouraged to think of electricity in the same way as he thinks of mass. The ampere is then introduced as a fourth basic quantity, and rationalized mksa units are used throughout the text.*

The third and fourth chapters continue with the discussion of fields by extending the circuit-theory laws for the capacitor and the inductor in the same way as Ohm's law was extended in Chapter 1. Much of the argument is carried

---

* For a thorough discussion of this subject, see Silsbee, *Systems of Electrical Units*, NBS Monograph No. 56, September, 1962.

out by analogy with Chapter 1. The electric field and flux density, the magnetic field and flux density, and permittivity and permeability are thereby introduced. The conventional laws of Gauss, Coulomb, Ampere, and Biot and Savart are deduced as consequences of the theory, and thus provide experimental confirmation. By developing the subject in this manner, it is based on the laws of the resistor, capacitor, and inductor, with which the student will be thoroughly familiar. Few students today ever see the classic experiments of Coulomb or Ampere demonstrated, and still fewer repeat them for themselves in the laboratory. However, all will have used the circuit elements and verified these laws many times in their own experience. Therefore they seem to provide a better starting point for the present purpose.

In Chapter 5 the classic equations of Maxwell are collected, along with the constitutive relations. When the boundary conditions are added, there is then sufficient information for a complete solution for all five field quantities. Chapter 6 gives some examples of the solution of these equations for the static case, and Chapter 7 some examples on the calculation of resistance, capacitance, and inductance. The ideas of polarization and magnetization are introduced in Chapter 8 as alternatives to permittivity and permeability, to take account of the effects of the medium. This is to lay the foundations for Chapter 12.

The subject matter covered in the first eight chapters is that least likely to arouse the enthusiasm of the student. A knowledge of it is nevertheless essential to the understanding of those topics which are currently of more interest. Therefore a concerted effort was made to ensure that the treatment of this part of the subject did not extend to any greater length than is really necessary. By this means, half the book was left to deal with topics which the electrical-engineering student is much more likely to find interesting.

Chapters 9 and 10 treat waves, radiation, and waveguides. Here an attempt is made to treat a few cases relatively thoroughly rather than cover a wide range of topics. The rest of the text deals with the interaction of fields with particles and is based on the following development. In Chapter 11 the particles are considered to be independent and the fields quasi-static. In the last section the effects of space charge are introduced—the removal of the assumption that the particles move independently of each other. In Chapter 12 the assumption of independence is reimposed, but rapidly varying fields are considered. This leads to a treatment of anisotropic media. In Chapter 13 space charge effects are taken into account in the presence of rapidly varying fields, leading to a treatment of waves on beams and the ideas of the traveling wave tube. Finally, in Chapter 14 bulk mechanical motion of the medium is allowed, introducing the laws of mechanics into the problem, in addition to the electromagnetic laws. Throughout these chapters an attempt is made to relate the theory being developed to devices of current interest in electrical-engineering research.

The relegation of the polar-coordinate systems to the Appendix is done deliberately. Their introduction in the main text is not only unnecessary, but is a distinct encumbrance to the student. His attention is inevitably distracted from the fundamental concepts, which it is the purpose of the text to convey to him, by the intricacies of the mechanical manipulation required. These are much greater in the case of the polar systems than in the Cartesian system. It proved not to be possible to avoid the use of polar coordinates completely, however; otherwise there would have been no need for the Appendix. Rather than simply list the formulas, some examples are given to show how they can be deduced from the general definitions given in the main text. This is intended to discourage the "handbook attitude."

The aim of the book is twofold. First, to develop, as rapidly as possible for the electrical-engineering student, a good working knowledge of electromagnetic theory based on a sound understanding of its foundations, by making the maximum possible use of the electrical-engineering subjects previously studied. Second, to arouse the interest of the student in the subject by showing at an early stage how the theory is applied to topics of current importance in electrical engineering. It is hoped that this might dispel to some extent the often held attitude that the subject is dull and uninteresting.

I wish to thank Dean Ronald E. Scott for encouraging me to undertake the writing of the present work and for many helpful discussions during its early stages. I am also grateful to my colleague, Professor Clinton G. Hollins, for checking the manuscript and for making numerous suggestions.

Boston, Mass. R.D.S.
February 1965

Contents

CHAPTER 1

The Field and its Properties

1–1 Introduction

The notion of a field occurs in many branches of the physical sciences. The electrical engineer is most concerned with the electric and magnetic fields, and these will form the subject of this text. The fundamental laws on which the theory of fields is based have been discovered by experiment in the laboratory. In spite of the apparent preponderance of mathematics, in this subject we must never lose sight of the fact that we are dealing with a physical science, and the supreme test of any result we may derive is whether or not it agrees with experiment.

By the very nature of the way such a subject develops there is no one set of experimental results that forms the one and only starting point. A postulate based on experiment is made, from which new results are predicted. These, when verified, can form a new starting point from which the original results can be derived and which form the basis for further predictions. One way in which we can select our starting point is to choose those results with which we have had some experience in the past and which command our confidence. Thus we choose here results that are related as closely as possible to circuit theory, a subject with which we are already familiar.

1–2 Relation to circuit theory

The variables with which we work in circuit theory are *voltage* and *current*. The hardware consists of *resistors*, *capacitors*, and *inductors*. We shall consider only resistors in this section, and the behavior is governed by the well-known relation $V = RI$, which is due to Ohm and was discovered by experiment. The whole effect of the hardware is *lumped* into the one parameter R, and a network of resistors is commonly referred to as a *lumped-parameter network*. To define such a network we need only know the values of the resistors and how they are connected. The physical positions of the resistors are of no consequence, and they can be moved relative to each other without affecting our results, provided we do not alter the connections. The currents flow in the resistors and the connecting wires (the branches), and voltages are measured between nodes.

Next we may consider a somewhat more general situation. Suppose our circuit consists of a piece of resistance wire whose resistance varies along its length. This would form a distributed parameter circuit. In this case we need to define the position of a point P on the wire, and we might introduce a length parameter measured along the wire from one of its ends as shown in Fig. 1–1. We would further have to define a resistance per unit length of the wire, R_l. Then the resistance of an element of length dl at the point P would be $R_l\,dl$. We must also remember that the value of R_l depends upon the position of P; i.e., it is a function of l, and so we would write this resistance as $R_l(l)\,dl$. The voltage across this element would be $IR_l(l)\,dl$, where I is the current flowing in the wire. The potential difference between two points P_1 and P_2, distant l_1 and l_2 from the end of the wire, is then given by the integral

$$V_2 - V_1 = \int_{l_1}^{l_2} IR_l(l)\,dl.$$

In this case the current is still constrained by the circuit, i.e. it must flow along the wire.

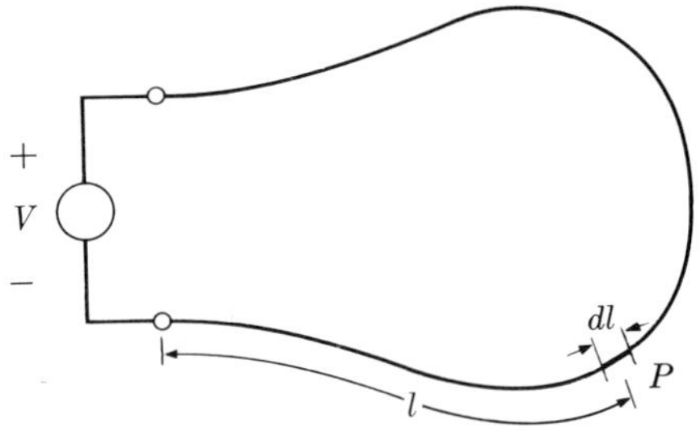

FIG. 1–1. A distributed circuit.

Let us generalize the situation still further by considering currents flowing in a three-dimensional conducting region. For example, we can think of electric currents flowing in the ocean (perhaps caused by the motion of the sea in the magnetic field of the earth). To define the position of a point in our "circuit," we need three coordinates instead of one as in the previous example. We can think of defining position in terms of Cartesian coordinates, which means specifying the three coordinates x, y, and z with reference to some suitable origin. Furthermore, the circuit no longer confines the current as it did in the previous examples. There is no wire at the point P to show us in which direction the current will flow. When we have determined the direction of flow of the current at a point either by calculation or by measurement, we can then visualize an arrow at this point to denote the information. We would also like to know the magnitude of the current. In the case of a circuit the current is zero outside the walls of the wire, and no difficulty is encountered; we measure the total current flowing through the total cross-sectional area of the wire. In this case, however, there is no such constraint, and the current we measure will depend upon how large a cross-sectional area we take. We might think of taking a unit area perpendicular to the direction of flow and measuring the current flowing through this. We need to refine this definition, however, to take account of the fact that the current flow may not be uniform across the area. We take the value of the current flow and divide by the area and then reduce the size of the area until in the limit it becomes zero, that is, we take

$$\lim_{\Delta S \to 0} \frac{I}{\Delta S},$$

which defines the current per unit area or the current density at the point. We now have a magnitude and a direction to provide us with information about the current density; i.e., the current density **J** is a vector.*

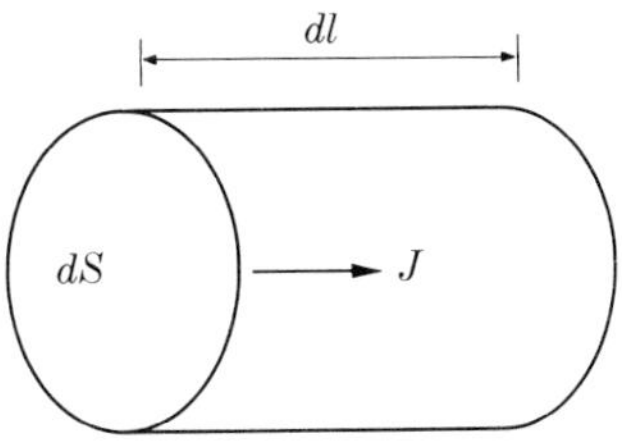

FIG. 1-2. An element of volume in the electric-current field.

If there are currents flowing and the medium is not a perfect conductor, then potential differences will be produced. Let us consider an element of volume with a cross section dS perpendicular to the current density and length dl (Fig. 1-2). Suppose the resistance of this element is R. Then the potential difference across the element will be $V = RJ\,dS$. The resistance, and therefore the voltage, will depend upon the magnitude of dl. However, since both the current and the resistance of the medium may vary from point to point, it is convenient to take a limit and so obtain a second quantity defined at the point. The limit we need is

$$\lim_{dl \to 0} \frac{V}{dl},$$

which defines a second magnitude denoted by E. This quantity also has a direction, namely the same direction as **J**. We now have the relation

$$\frac{V}{dl} = E = \frac{R\,dS}{dl} J.$$

The quantity $(1/R)\,(dl/dS)$ defines the conductivity σ of the medium, and **E** is known as the *electric field*. Thus, bearing in mind that **E** and **J** are in the same direction, we can write

$$\mathbf{J} = \sigma \mathbf{E}.$$

This is the field form of the circuit relation $I = GV$, where G is the conductance. The equation illustrates the existence of two characteristic types of vectors which it is convenient to distinguish. The vector **J** represents the flow of the current

* The notation used in this text for vectors and a brief review of the operations of vector algebra is given in the Appendix.

and may be thought of as a *flow vector* or a *flux vector*. The vector **E** may be thought of as the forcing term which produces the current and constitutes the *force vector*.

We have seen how the quantities **E** and **J** are vectors defined at a point. If at each point in space we define a vector, we have defined a *vector field* or a *vector point function*. The quantity σ, which is a scalar, may also vary with position in the medium, and if we define a scalar quantity at each point in space, we have a *scalar field* or a *scalar point function*.

We can represent a vector field graphically by drawing a curve in space, so that at each point the tangent to the curve represents the direction of the vector. For a quantity such as **J** these lines are *flux lines*, and for quantities such as **E** they are *lines of force*. Faraday made extensive use of the concept of these lines, and much subsequent work has been based on them. These lines also convey information about the magnitude of the vector as well as its direction. We can see that as the flux lines converge, a greater quantity is flowing through a given cross-sectional area, and hence the flux vector is greater.

1-3 Equipotentials and the gradient

We saw in the last section how an element of the medium can have a potential difference across it. If we arbitrarily assign the value of zero potential to some point in the medium, then we can define a potential at each point in the medium. Since potential is a scalar quantity, we now have a scalar field. If we take all the points in the field which have a given fixed value of potential, we obtain a surface. For each given value of the parameter, potential, there will be a different surface. These surfaces are known as *equipotential surfaces*.

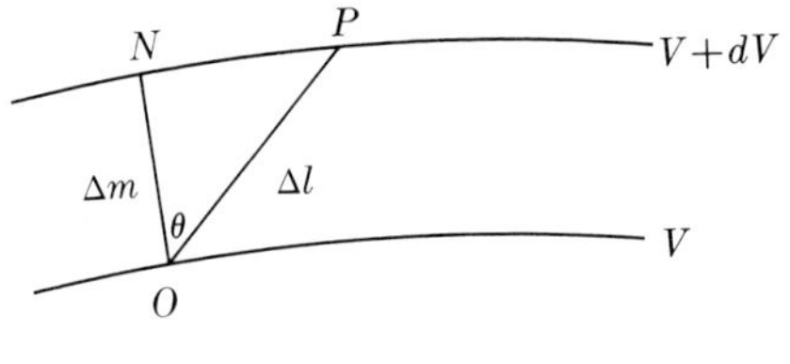

FIG. 1-3. Equipotential surfaces and gradient.

If we move from one equipotential surface to another, then by definition the change in potential must always be the same no matter what route we take from one surface to the other. Let us consider the two equipotential surfaces shown in cross section in Fig. 1-3, with potentials V and $V + \Delta V$. If we start at the point O and move to the point P, then we experience a change in potential, ΔV. If we let the length of OP be Δl, we can define a derivative as we are accustomed to doing in calculus, as follows:

$$\lim_{\Delta l \to 0} \frac{V(P) - V(O)}{\Delta l} = \lim_{\Delta l \to 0} \frac{\Delta V}{\Delta l} = \frac{dV}{dl}.$$

Clearly the magnitude of this limit will depend upon the direction in which we go from one surface to the other, and we therefore have the idea of a directional derivative. The shortest distance between the two surfaces is, of course, the perpendicular distance ON. This then will give us the maximum value of the derivative. This special value is known as the *gradient* of the scalar field and is written grad V. As we have seen, this derivative is associated with a particular direction, and is therefore a vector. If we take the gradient of a scalar field, we obtain a vector field. From Fig. 1–3 we see that $\Delta m = \Delta l \cos\theta$. Hence the directional derivative in the OP-direction is given by

$$\lim \frac{\Delta V}{\Delta l} = \lim \frac{\Delta V}{\Delta m}\cos\theta = \text{grad } V \cos\theta.$$

We shall define directions by unit vectors, for example, $\mathbf{u}_l$ is a vector of unit magnitude in the l-direction (OP). The directional derivative in the $\mathbf{u}_l$-direction is the component of the gradient in this direction. Using vector algebra we can write this as

$$\text{grad}_l V = |\text{grad } V| \cos\theta = \text{grad } V \cdot \mathbf{u}_l,$$

where $|\text{grad } V|$ is the magnitude of the vector grad V.

It should be pointed out here that $\text{grad}_l\, V$ is a scalar quantity representing the magnitude of the derivative in the $\mathbf{u}_l$-direction, but grad V is a vector containing both a magnitude and a direction (in this case, the ON-direction) as discussed above.

We can now write the gradient in terms of the Cartesian coordinate system. From the discussion above we see that

$$\text{grad}_x\, V = \text{grad } V \cdot \mathbf{u}_x = \frac{\partial V}{\partial x}\cdot$$

We can therefore write the gradient as follows:

$$\text{grad } V = \mathbf{u}_x \frac{\partial V}{\partial x} + \mathbf{u}_y \frac{\partial V}{\partial y} + \mathbf{u}_z \frac{\partial V}{\partial z}\cdot$$

It is convenient at this point to introduce the commonly used operational notation. We may write

$$\mathbf{u}_x \frac{\partial V}{\partial x} + \mathbf{u}_y \frac{\partial V}{\partial y} + \mathbf{u}_z \frac{\partial V}{\partial z} = \left(\mathbf{u}_x \frac{\partial}{\partial x} + \mathbf{u}_y \frac{\partial}{\partial y} + \mathbf{u}_z \frac{\partial}{\partial z}\right) V,$$

and define the quantity in parentheses as the vector operator ∇ (del), that is,

$$\nabla = \mathbf{u}_x \frac{\partial}{\partial x} + \mathbf{u}_y \frac{\partial}{\partial y} + \mathbf{u}_z \frac{\partial}{\partial z}\cdot$$

It should be noted that ∇ has the properties of both a vector, because it contains the vectors $\mathbf{u}_x$, $\mathbf{u}_y$, $\mathbf{u}_z$, and an operator, because it contains the differential operators $\partial/\partial x$, $\partial/\partial y$, $\partial/\partial z$. We may now write the gradient as ∇V. This should be regarded merely as a shorthand notation for the above expression.

Referring back to our definition of the electric field **E**, we see that it is closely related to grad V. In fact, the two are equal in magnitude. Since we are now using vector notation, we must be careful to get the sense of the vectors correct. Let us consider a current flowing in the positive x-direction. Since current flows from points of higher to points of lower potential, the potential must then decrease as x increases. The electric field **E** is in the same direction as the current, and therefore, if we consider two points A and B at a distance x and $x + \Delta x$ from the origin, then the potential of B is lower than that of A. Thus

$$E = \lim_{\Delta x \to 0} \frac{V(B) - V(A)}{\Delta x} = \lim_{\Delta x \to 0} \frac{(V - \Delta V) - V}{\Delta x} = -\frac{dV}{dx},$$

or, in general,

$$\mathbf{E} = -\text{grad } V.$$

This expression enables us to obtain **E** when V is given. The inverse operation is also of interest. Consider again Fig. 1-3. The potential change in moving from O to P is ΔV, and may be written

$$\Delta V = \frac{\Delta V}{\Delta l} \Delta l = \frac{\Delta V}{\Delta m} \Delta l \cos \theta,$$

which in the limit becomes grad $V \cdot d\mathbf{l}$ or $-\mathbf{E} \cdot d\mathbf{l}$. Any arbitrary path can be made up of a number of vector elements of length $d\mathbf{l}$, and the potential change between any two points can therefore be obtained by summing the potential changes over all the line elements. Thus, for the potential difference between two points P_1 and P_2, we have $V_1 - V_2 = \int_1^2 \mathbf{E} \cdot d\mathbf{l}$.

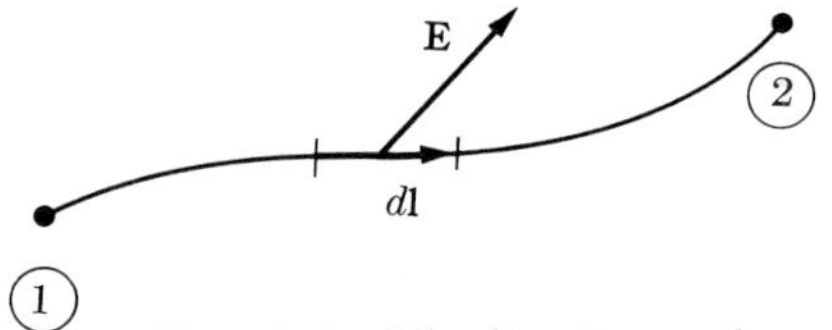

FIG. 1-4. The line integral.

The right-hand side of this equation is a *line integral*. Its meaning can now be simply explained. We take two points in a vector field and join them by a line (not necessarily straight). At each point on this line we have two vectors: **E** defined by the vector field, and $d\mathbf{l}$ of infinitesimal length defining the direction of the line at that point (Fig. 1-4). We now form the scalar product, and add the contributions from all elements which make up the line. This is a general definition of a line integral.

The particular vector field **E** considered here belongs to a special class of vector fields which have the property that they can be expressed as the gradient of a scalar field (we had $\mathbf{E} = -\text{grad } V$). Not all vector fields have this property, as we shall see later. An important property of this class of fields is that the value of their line integral depends only upon the end points and not on the route taken between them. We have already seen this, but it may readily be demonstrated as follows:

$$\int_1^2 \mathbf{E} \cdot d\mathbf{l} = \int_1^2 (-\text{grad } V) \cdot d\mathbf{l} = -\int_1^2 \text{grad}_l\, V\, dl = -\int_1^2 \frac{\partial V}{\partial l}\, dl$$
$$= -\int_1^2 dV = V_1 - V_2.$$

Thus the integral depends only upon the potential at the end points and is independent of the path. If the integral is taken around a closed path so that the end point coincides with the beginning of the line, then the result must be zero, and we write

$$\oint \text{grad } V \cdot d\mathbf{l} = 0.$$

The class of vector fields under discussion here is known as *conservative.*

1–4 Sources and divergence

We have considered above currents flowing in a three-dimensional continuum, such as the ocean, but so far have paid no attention to how such currents originate. The idea of sources in the most general sense in field theory is a rather abstract one, but we shall approach it here from a physical picture. Let us imagine an insulated copper wire inserted into our ocean and bared at the tip, the other end being connected to a battery. The other side of the battery can be imagined to be connected to a large copper box which the ocean fills. We can now envisage current flowing away from the tip of the wire, and so constitute a source of current. We do not want to disturb the system in any way by the mechanism by which we produce the source, so that in order to leave the flow unchanged by the wire we must imagine the wire to have zero dimensions. It would be even better to go one step further and take away the wire altogether, and merely define in a theoretical fashion a source at the point so that the resulting effect is the same.

A point source thus constitutes a singular point in the field, and the current flows away from it in all directions. Representing the field graphically in terms of flux lines, we can think of a source as a point at which flux lines originate. In general, we would allow both positive and negative sources, and we could regard a negative source as a point at which flux lines terminate. This is sometimes referred to as a *sink.* We can extend this idea to include a continuous distribution of sources throughout the whole volume of the field. If we consider a volume within the field, then it is seen that the net outflow of current across the surface of this volume is dependent upon the sources inside the volume. We now wish to express this idea mathematically.

If we have an area ΔS perpendicular to the current vector $\mathbf{J}$, then the current flowing through this area is $J\Delta S$. If the normal to the surface ΔS makes an angle θ with the vector $\mathbf{J}$ (Fig. 1-5), then the current is $J\Delta S \cos\theta$ since $\Delta S \cos\theta$ is the projected area perpendicular to the current flow. It is convenient to regard the area as a vector, its direction being the normal to the surface. We can then write the current as $\mathbf{J} \cdot \mathbf{\Delta S}$. It may be noted that this definition of vector area leaves the sense undetermined. If $\mathbf{\Delta S}$ forms part of a larger surface which is closed, then the outward-drawn normal is conventionally considered to be positive. Let us consider a volume within the field, and let the surface enclosing it be made up of many such elements of area $\mathbf{\Delta S}$. We can then obtain the net current flowing out of this volume by calculating $\oint\!\!\oint \mathbf{J} \cdot d\mathbf{S}$, the circle on the integral sign indicating that the surface is closed. Note that the direction of $d\mathbf{S}$ is positive by definition. If the current at this point is flowing outward, then the scalar product $\mathbf{J} \cdot d\mathbf{S}$ is also positive. If the flow is inward, the scalar product is negative. The integral therefore represents the net outward flow of current. Its value will vary with its position in the field, and, following the procedure already developed, we think of shrinking the volume down to zero and so obtain the value of the integral at a point. However, if the volume goes to zero, then so does its surface area, and if the field is finite, then the value of the integral will also go to zero. We can overcome this difficulty by taking the following limit and thereby obtain a definition of *divergence*:

$$\operatorname{div} \mathbf{J} = \lim_{\Delta\tau \to 0} \frac{\oint\!\!\oint \mathbf{J} \cdot d\mathbf{S}}{\Delta\tau},$$

where $\Delta\tau$ is the volume of the element, and the integral is taken over the surface enclosing it. The choice of the term "divergence" becomes obvious in the light of the earlier discussion.

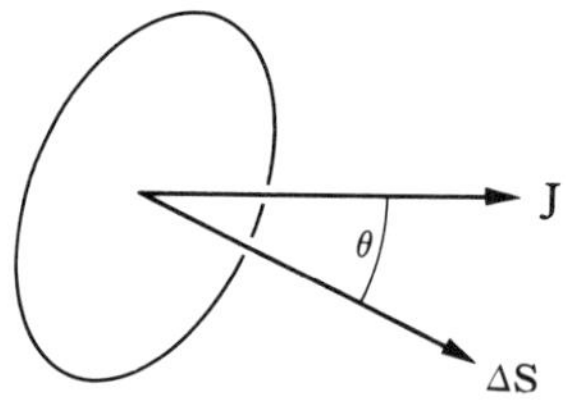

FIG. 1-5. The use of vector area to obtain current.

A better insight into this expression can be gained by carrying out the calculation in Cartesian coordinates. Consider the volume element $\Delta x\, \Delta y\, \Delta z$ shown in Fig. 1-6. Let us first calculate the flow across the horizontal faces. The area of the bottom face is $\Delta x\, \Delta y$, and the outward-drawn normal is in the negative z-direction. Thus $\mathbf{\Delta S} = -\mathbf{u}_z\, \Delta x\, \Delta y$. The flow over this face is therefore given by $-\Delta x\, \Delta y\, \mathbf{u}_z \cdot \mathbf{J} = -J_z\, \Delta x\, \Delta y$. Similarly for the top face, $\mathbf{\Delta S}$ is given by $+\mathbf{u}_z\, \Delta x\, \Delta y$, and the flow is the value of the J_z-component over this face times the area $\Delta x\, \Delta y$. However, since $\mathbf{J}$ is a function of position, the value of its z-component for the two faces will be different, and if J_z is the value for the bottom face, then we may write its value for the top face as $J_z + (\partial J_z/\partial z)\, \Delta z$.

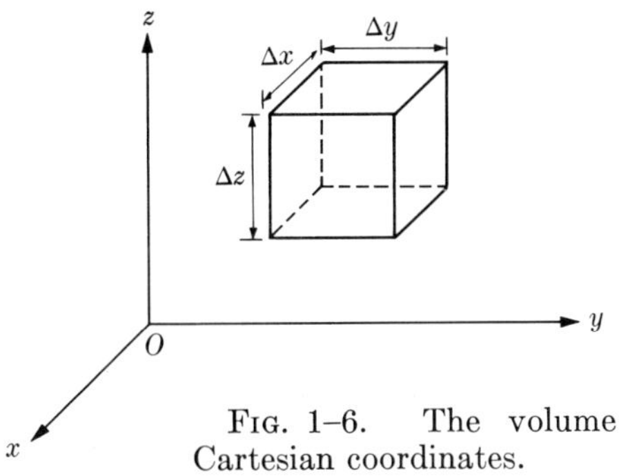

FIG. 1–6. The volume element in Cartesian coordinates.

Thus the net outward flow becomes

$$-J_z\,\Delta x\,\Delta y + \left(J_z + \frac{\partial J_z}{\partial z}\,\Delta z\right)\Delta x\,\Delta y = \frac{\partial J_z}{\partial z}\,\Delta x\,\Delta y\,\Delta z.$$

Similarly the net outward flow over the other two pairs of faces can be calculated, and the total outward flow is given by

$$\left(\frac{\partial J_x}{\partial x} + \frac{\partial J_y}{\partial y} + \frac{\partial J_z}{\partial z}\right)\Delta x\,\Delta y\,\Delta z.$$

But $\Delta x\,\Delta y\,\Delta z$ is the volume $\Delta\tau$ of the element. Thus for the limit defining div **J** we have

$$\operatorname{div}\mathbf{J} = \frac{\partial J_x}{\partial x} + \frac{\partial J_y}{\partial y} + \frac{\partial J_z}{\partial z}.$$

It is readily verified that the divergence can be written in terms of the del-operator as a scalar product:

$$\operatorname{div}\mathbf{J} = \nabla\cdot\mathbf{J}.$$

We shall now derive mathematically a theorem known as the *divergence theorem*. Let us consider the volume element $\Delta\tau_1$ (Fig. 1–7) and calculate the net outward flow per unit volume:

$$\phi_1 = \frac{\oint\!\!\oint_{\Delta\tau_1}\mathbf{J}\cdot d\mathbf{S}}{\Delta\tau_1}.$$

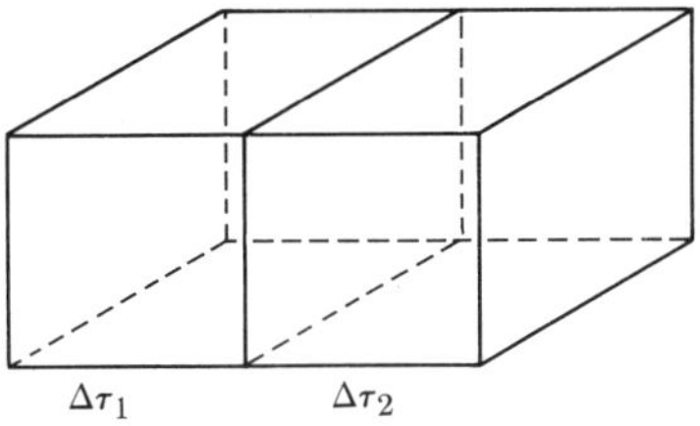

FIG. 1–7. Two contiguous volume elements.

Now let us take a contiguous volume element $\Delta\tau_2$ with a common face, calculate the net outward flow per unit volume, and add, obtaining the result

$$\phi_1\,\Delta\tau_1 + \phi_2\,\Delta\tau_2 = \oint\!\!\oint_{\Delta\tau_1}\mathbf{J}\cdot d\mathbf{S} + \oint\!\!\oint_{\Delta\tau_2}\mathbf{J}\cdot d\mathbf{S}.$$

Here the surface integral over the common face occurs twice, but with opposite sign since the outward-drawn normals for this face are in opposite senses for the two elements $\Delta\tau_1$ and $\Delta\tau_2$. Thus the right-hand side of the equation is equivalent to the surface integral $\oiint \mathbf{J} \cdot d\mathbf{S}$ taken over the surface enclosing the whole volume of $\Delta\tau_1$ and $\Delta\tau_2$. Alternatively, we can say that what flows out of $\Delta\tau_1$ at the common interface must flow into $\Delta\tau_2$, and therefore the two contributions cancel. We can now add more volume elements so as to build up any given arbitrary volume V. The left-hand side will then be a sum of $\phi\,\Delta\tau$ over all the volume elements throughout V, and by an extension of the argument above, the right-hand side will be the surface integral taken over the surface S enclosing V. Thus

$$\sum_V \phi\,\Delta\tau = \oiint_S \mathbf{J} \cdot d\mathbf{S}.$$

If we now make $\Delta\tau \to 0$, then ϕ becomes div $\mathbf{J}$ and the summation becomes an integral. Hence we have

$$\iiint_V \operatorname{div} \mathbf{J}\, d\tau = \oiint_S \mathbf{J} \cdot d\mathbf{S},$$

where S is the surface enclosing V. This is the divergence theorem. It is, of course, quite general, and $\mathbf{J}$ could represent any vector field.

1-5 The equation of continuity

One of the fundamental laws of this subject is the law of conservation of charge, which states that electric charge can neither be created nor destroyed. Its truth must be verified by experience. The equation of continuity is a mathematical formulation of this law.

Consider again the physical picture of sources discussed in the last section, but instead of connecting the wire to a battery assume it to be connected to a capacitor. Charge flowing out of the capacitor would then constitute the current. We must now extend this idea to the abstract notion of a capacitor of zero dimensions situated at a point, so that charge can be stored or can build up at a point. If we now consider a volume element and apply the law of conservation of charge, we see that any net outflow of charge must be accompanied by a decrease in the charge on the capacitor. The naive idea of storing charge at a point is made more reasonable by considering charge density. This is obtained by measuring the charge in a given volume and dividing by the volume. However, since the charge may not be uniformly distributed, we take a very small volume and in the limit make it vanishingly small. We could thus define the charge density q_v at a point as follows:

$$q_v = \lim_{\Delta\tau \to 0} \frac{Q}{\Delta\tau},$$

where Q is the charge contained in $\Delta\tau$. We can now think of q_v increasing and decreasing.

If we consider a volume element $\Delta\tau$, then the charge contained within it is given by $q_v\,\Delta\tau$. By the law of conservation of charge, the time rate of decrease of this integral must be equal to the net outflow from $\Delta\tau$. But the net outflow is given by $\oint\!\!\oint \mathbf{J}\cdot d\mathbf{S}$. Hence

$$\oint\!\!\oint \mathbf{J}\cdot d\mathbf{S} = -\frac{\partial}{\partial t} q_v\,\Delta\tau.$$

We next divide through by $\Delta\tau$ and take the limit as $\Delta\tau \to 0$. The left-hand side then becomes div $\mathbf{J}$, and we have

$$\operatorname{div}\mathbf{J} = -\frac{\partial q_v}{\partial t},$$

which is the equation of continuity.

If we have a vector field $\mathbf{J}$ related to a scalar field s by the relation div $\mathbf{J} = s$, then s is referred to as the *source* of the vector field. Thus $-(\partial q_v/\partial t)$ is the source of $\mathbf{J}$.

1–6 The curl of a vector field

Kirchhoff's second law in circuit theory states that the sum of the voltages around a closed loop is equal to zero. Thus if we consider the distributed circuit in Fig. 1–8, then by Kirchhoff's law,

$$-\int_A^B E\,dl + V = 0.$$

If there is no battery in the circuit, then $V = 0$, and $\oint E\,dl = 0$. (Compare this with the result $\oint \mathbf{E}\cdot d\mathbf{l} = 0$ for fields.) On this basis, if there is no battery in the circuit, no current can flow. However, this picture is incomplete. We can produce a current in a closed loop of wire by means of electromagnetic induction. If there is a current in the wire and the wire has a finite conductivity, then we must have the result $\oint E\,dl \neq 0$.

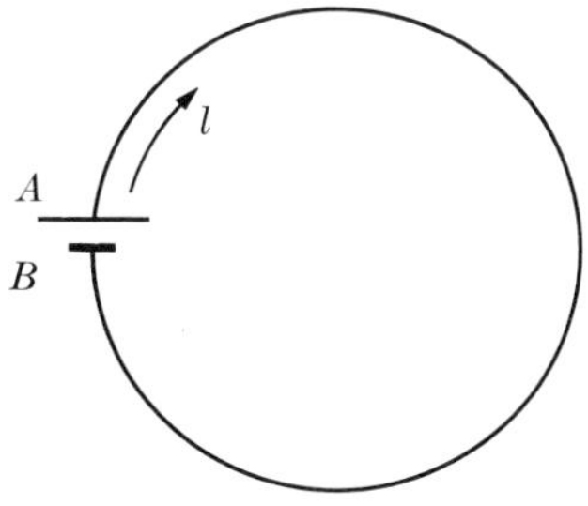

FIG. 1–8. A single-loop distributed circuit.

If we return to our picture of currents flowing in the ocean, we can visualize such current whirls or vortices flowing here too. These must now have the property that $\oint \mathbf{E}\cdot d\mathbf{l} \neq 0$. This forms the basis of a new parameter which we can define. Let us take any closed curve in the field. (For simplicity we will think of this curve as being described in a plane.) We now calculate $\oint \mathbf{E} \cdot d\mathbf{l}$ around the loop. This will vary from point to point, and therefore we will again want to make the loop small and in the limit vanish, so that we can define a quantity at a point. However, again we run into the problem that if we make the loop of zero dimensions and the field is finite, then the integral will vanish. We can overcome this difficulty by dividing the integral by the area enclosed by the loop and taking the limit as the area goes to zero. Another complication arises in this case, however. The loop can be orientated in different directions, and we may expect the result to depend upon this. In order to define a quantity uniquely, we calculate this limit for all orientations of the loop and take the maximum value. This defines a magnitude and a direction, the direction being that direction of the vector area enclosed by the loop which produces the maximum value of the integral. Thus the quantity defined is a vector. We still have to define the sense of this vector, which we do by saying that the direction is related to the direction around the loop in which the integral is taken by the right-hand screw rule (Fig. 1–9). The vector so defined is known as the *curl* of the vector field. Hence

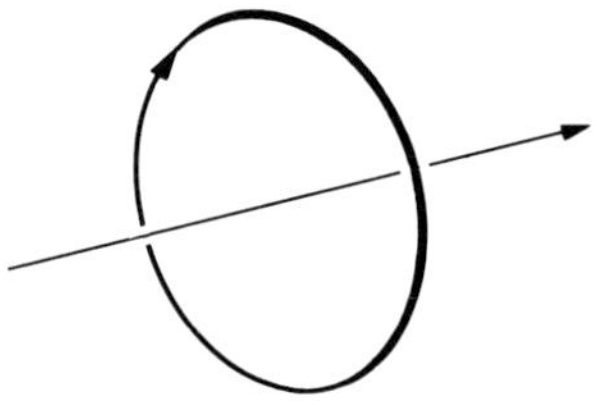

FIG. 1–9. The right-hand screw rule.

$$\operatorname{curl} \mathbf{E} = \lim_{\Delta S \to 0} \frac{(\oint \mathbf{E} \cdot d\mathbf{l})_{\max}}{\Delta S} \mathbf{u}_m.$$

We shall now calculate the curl in Cartesian components. We calculate the three components of the vector and then add vectorially rather than going through a maximization procedure. To obtain the x-component we use the element of area $\Delta y\, \Delta z$. The direction of integration around the rectangle must be related to the x-direction by the right-hand rule. For a right-handed set of axes, the x-axis will be down into the page (Fig. 1–10), and hence the direction of integration is as shown. We must now calculate $\oint \mathbf{E} \cdot d\mathbf{l}$ around this rectangle. If we start at the corner A, then $d\mathbf{l}$ for the first side is $\mathbf{u}_y\, \Delta y$, and $\mathbf{E} \cdot d\mathbf{l}$ is $\mathbf{E} \cdot \mathbf{u}_y\, \Delta y = E_y\, \Delta y$. For the side CD, $d\mathbf{l}$ is $-\mathbf{u}_y\, \Delta y$ and the y-component

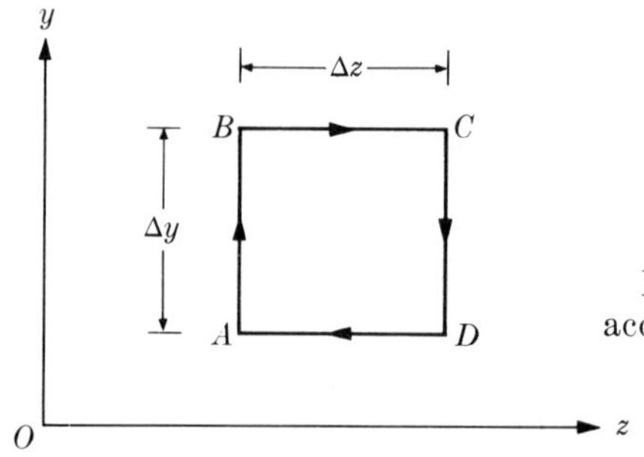

FIG. 1–10. An integral taken in accordance with the right-hand rule.

of **E** will be given by $E_y + (\partial E_y/\partial z)\,\Delta z$. The contribution to the line integral from these two sides is therefore given by

$$E_y\,\Delta y - \left(E_y + \frac{\partial E_y}{\partial z}\,\Delta z\right)\Delta y = -\frac{\partial E_y}{\partial z}\,\Delta y\,\Delta z.$$

Similarly the contribution from the sides BC and DA is given by

$$\left(E_z + \frac{\partial E_z}{\partial y}\,\Delta y\right)\Delta z - E_z\,\Delta z = \frac{\partial E_z}{\partial y}\,\Delta y\,\Delta z,$$

and hence

$$\oint \mathbf{E}\cdot d\mathbf{l} = \left(\frac{\partial E_z}{\partial y} - \frac{\partial E_y}{\partial z}\right)\Delta y\,\Delta z.$$

But $\Delta y\,\Delta z$ is the area ΔS enclosed by the loop, and so

$$\lim_{\Delta S\to 0}\frac{\oint \mathbf{E}\cdot d\mathbf{l}}{\Delta S} = \frac{\partial E_z}{\partial y} - \frac{\partial E_y}{\partial z}\cdot$$

This is the x-component of curl **E**. Similarly the y- and z-components can be calculated, and we have finally for curl **E**,

$$\text{curl}\,\mathbf{E} = \left(\frac{\partial E_z}{\partial y} - \frac{\partial E_y}{\partial z}\right)\mathbf{u}_x + \left(\frac{\partial E_x}{\partial z} - \frac{\partial E_z}{\partial x}\right)\mathbf{u}_y + \left(\frac{\partial E_y}{\partial x} - \frac{\partial E_x}{\partial y}\right)\mathbf{u}_z.$$

Examination of this expression shows that it may be written in terms of the del-operator as a cross product,

$$\text{curl}\,\mathbf{E} = \nabla\times\mathbf{E},$$

or it may be conveniently remembered by the determinant

$$\text{curl}\,\mathbf{E} = \begin{vmatrix}\mathbf{u}_x & \mathbf{u}_y & \mathbf{u}_z\\ \partial/\partial x & \partial/\partial y & \partial/\partial z\\ E_x & E_y & E_z\end{vmatrix}\cdot$$

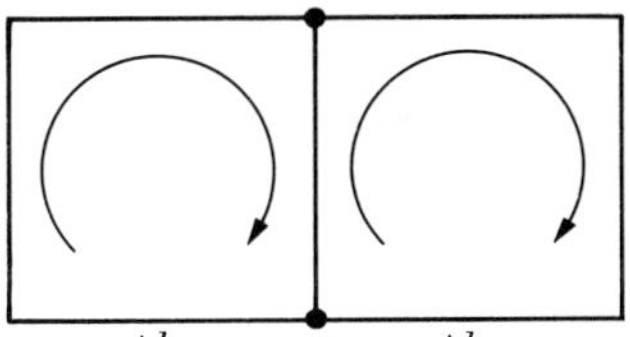

FIG. 1–11. Two contiguous loops.

We may now derive another mathematical theorem which is due to Stokes. Let us consider two contiguous loops with a common side as shown in Fig. 1–11. If we calculate the integral $\oint \mathbf{E}\cdot d\mathbf{l}$ taken around Δl_1, and add to it the integral taken around Δl_2 (the direction of integration around the loops being in the same sense for both cases as shown in the figure), it is seen that the common side is traversed twice, once in each direction. These two contributions to the sum will therefore be equal and opposite and so will cancel. We are then left with the integral taken around the periphery. We next define a vector **A** by the relation

$$\mathbf{A} = \frac{(\oint \mathbf{E} \cdot d\mathbf{l})_{\max}}{\Delta S} \mathbf{u}_m.$$

Thus **A** is defined so that the limit of **A** as ΔS tends to zero is curl **E**. If $\mathbf{\Delta S}_1$ is the vector area enclosed by Δl_1, then the integral calculated above is the component of $\mathbf{A}_1$ in the $\mathbf{\Delta S}_1$-direction times the magnitude of $\mathbf{\Delta S}_1$, that is,

$$\mathbf{A}_1 \cdot \mathbf{\Delta S}_1 = \oint_{\Delta l_1} \mathbf{E} \cdot d\mathbf{l},$$

where $\mathbf{\Delta S}_1$ is supposed to be small enough to consider **A** to have the constant value $\mathbf{A}_1$. The sum is

$$\mathbf{A}_1 \cdot \mathbf{\Delta S}_1 + \mathbf{A}_2 \cdot \mathbf{\Delta S}_2 = \oint \mathbf{E} \cdot d\mathbf{l},$$

where the integral is taken around the loop enclosing both ΔS_1 and ΔS_2. We can extend this argument and build up any curve as the sum of many elements such as ΔS_1 and ΔS_2. The left-hand side then becomes the sum of all the $\mathbf{A} \cdot \mathbf{\Delta S}$ contributions summed over the surface, and the integral is taken around the curve bounding the surface:

$$\sum_s \mathbf{A} \cdot \mathbf{\Delta S} = \oint \mathbf{E} \cdot d\mathbf{l}.$$

Finally, taking the limit as $\Delta S \to 0$, we have

$$\iint \operatorname{curl} \mathbf{E} \cdot d\mathbf{S} = \oint \mathbf{E} \cdot d\mathbf{l}.$$

This is Stokes' theorem. It should be noted that the line integral is taken around a closed contour, and the surface integral may be taken over *any* surface which is bounded by that contour. The direction of integration around the closed contour is related to the direction of the vector area $d\mathbf{S}$ by the right-hand rule.

We obtained the type of field discussed in this section by considering the currents produced by electromagnetic induction. We shall not pursue the idea of the source of this type of field at this point. It is worth noting, however, that in our example we would need a time-varying magnetic field to produce this field. For the static (i.e., time-independent) case, we have the relation $\oint \mathbf{E} \cdot d\mathbf{l} = 0$, which is familiar in electrostatics.

1-7 Some special cases of vector fields

We saw that we can derive a vector field from a scalar field by taking the gradient. Such a field is known as a *scalar potential field*. We also saw that if $\mathbf{E} = -\operatorname{grad} V$, then $\oint \mathbf{E} \cdot d\mathbf{l} = 0$. Now

$$\operatorname{curl} \mathbf{E} = \lim_{\Delta S \to 0} \frac{(\oint \mathbf{E} \cdot d\mathbf{l})_{\max}}{\Delta S} \mathbf{u}_m = 0.$$

Thus the curl of a scalar potential field is everywhere zero. This type of field, which is often referred to as an *irrotational field*, also gives us the vector identity

$$\text{curl grad} = 0.$$

We might also consider the possibility of deriving a vector field as the curl of another vector field:

$$\mathbf{B} = \text{curl } \mathbf{A}.$$

This gives rise to another special class of fields known as *vector potential fields*. Let us now calculate the divergence of such a field:

$$\text{div } \mathbf{B} = \text{div curl } \mathbf{A} = \lim_{\Delta\tau\to 0} \frac{\oiint \text{curl } \mathbf{A} \cdot d\mathbf{S}}{\Delta\tau}.$$

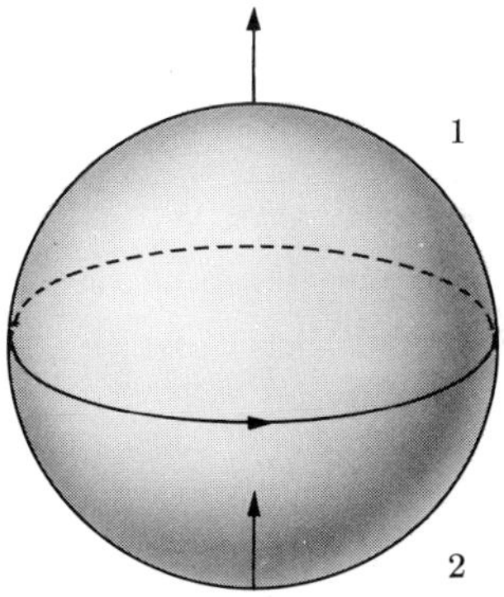

FIG. 1–12. A closed surface divided into two by a closed contour.

The integral is taken over a closed surface. Let us draw any closed contour on this surface, thereby dividing it into two surfaces as shown in Fig. 1–12. If we consider the surface integral over the upper surface 1, then by Stokes' theorem this integral is equal to $\oint \mathbf{A} \cdot d\mathbf{l}$ around the closed contour, the direction of flow being outward across the surface for the direction of integration around the contour shown (related by the right-hand rule). The line integral is also equal to the flow across surface 2, by Stokes' theorem, the direction of flow again being related by the right-hand rule and therefore, in this case, representing inward flow, as may be seen from the figure. Thus the flow over surfaces 1 and 2 cancel, and the net outward flow is zero, and therefore

$$\oiint \text{curl } \mathbf{A} \cdot d\mathbf{S} = 0,$$

from which we immediately have div $\mathbf{B} = 0$. Thus the divergence of a vector potential field is everywhere zero. This type of field is sometimes referred to as *solenoidal*. We also obtain another vector identity, namely,

$$\text{div curl} = 0.$$

The general vector field, one which has both curl and divergence, can be obtained as the superposition of an irrotational and a solenoidal field.

Problems

1–1 By considering the appropriate limit calculate the gradient of the function $f(r) = 1/r$, where r is the distance from the origin.

1–2 If the equipotential surfaces are planes parallel to the yz-plane, what is the direction of the gradient? If the potential varies inversely with the distance from the yz-plane, what is the magnitude of the gradient? Discuss the behavior at $x = 0$.

1–3 The potential is given by $V = -xy^2z$; find the electric field. In which direction is the field in (a) the xy-plane, (b) the yz-plane? Note that the y-axis is in both of these planes. Comment on your results.

1–4 Given that the electric field varies inversely with the distance from the yz-plane and is in the x-direction, calculate the potential difference between two points distant a and b from the plane.

1–5 The electric field varies inversely with the square of the distance from the origin and is directed radically outward. Show how to set up a line integral to calculate the potential difference between the points $(b, 0)$ and $(0, a)$.

1–6 Use the definition of divergence to calculate the divergence of a vector field directed radially outward from the origin, with magnitude $1/r^2$. (Consider a volume element consisting of a truncated cone with apex at the origin.)

1–7 Is the result of Problem 1–6 valid at the origin? If not, explain why, and then set up a suitable volume element to calculate its value.

1–8 Given that $\mathbf{J} = x^3\mathbf{u}_x + xy^2\mathbf{u}_y + xyz\mathbf{u}_z$, calculate the current flowing out of a unit cube with its opposite corners at the origin and the point $(1, 1, 1)$. Obtain the result (a) directly, and (b) using the divergence theorem.

1–9 The equipotential surfaces consist of concentric cylinders centered on the y-axis, and the potential is inversely proportional to the distance from the axis. Determine the electric charge density as a function of position and time, given that the medium has conductivity σ. [*Note:* This problem should be worked from first principles, using the definitions given in the chapter and not by the use of "standard formulas."]

1–10 Using the basic definition of curl, calculate the curl of the field of Problem 1–9.

1–11 For the electric field given by $\mathbf{E} = x^2(1 + y)\mathbf{u}_x + 2xy^2\mathbf{u}_y + z^2(1 + x)\mathbf{u}_z$, find the line integral of $\mathbf{E}$ taken around the square with opposite corners at the origin and the point $(1, 1, 0)$. Verify your result by integrating curl $\mathbf{E}$ over the plane surface bounded by the square.

1–12 Calculate the curl of the vector field derived in Problem 1–1 from the basic definition. Is this result valid at the origin?

1–13 If the field lines consist of circles centered on the x-axis, what is the curl of the field? What is its value on the x-axis?

1–14 Which of the fields of the previous problems can be classified as irrotational or solenoidal?

CHAPTER 2

Fundamental Quantities and Units

2-1 Introduction

A physical theory cannot be put forward until considerable experimental evidence has been made available. The theory is then concerned with the magnitudes of the different quantities which can be observed and the relation between them. The observed quantities will have a magnitude and a physical nature. The physical nature of the quantity will be formulated in the light of experience and will be given a name, such as mass, force, electric current, etc. The magnitude is a number which denotes how large the effect is. In order to measure this, we need a unit for each type of physical quantity.

It is clearly undesirable to have an arbitrary unit for every physical quantity we introduce, and the interrelationships between the quantities make this unnecessary. For example, we define unit velocity as unit distance per unit time. However, there is a certain minimum number of physical quantities whose units we are compelled to define arbitrarily. In mechanics, experience has taught us that this number is three, the usual choice being mass, length, and time. The unit of *mass* is defined as the mass of a particular block of material carefully preserved at one of the standards laboratories. The unit of *length* is the distance between two scratches on a bar, and the unit of *time* is defined in terms of the rotation of the earth on its axis. These quantities are the fundamental quantities, and all other quantities are derived quantities which can be obtained from them.

It is easy to define the derived quantities. For example, we define force as mass multiplied by acceleration, acceleration being readily defined in terms of distance and time. The unit too is readily obtained, unit force being that required to give a unit mass unit acceleration. For the fundamental units, however, the situation is different. As we have already seen, the unit is defined arbitrarily. To define the quantity itself is no small matter. A moment's reflection will show that this is a point which is just passed over in elementary mechanics courses. For our present purposes we can regard the situation as follows. The fundamental quantities are our starting point, and we shall not attempt to define them. We *know* what they are in the sense that mass, length, and time convey a notion common to all of us. We know by direct experience in the laboratory what their manifestations are, and we have evolved suitable methods of manipulating them. (For example, we know that if we push against an object it will resist movement, and we compare two masses by means of a balance.)

2-2 Electrical quantities

Just as experience taught us that three fundamental quantities were sufficient in mechanics, it also teaches us that four are required if electric and magnetic quantities are involved. However, the extension has not been straightforward. Much has been written on the subject, and a number of different systems of units has come into use.

In electrical engineering the most common choice of the fourth quantity is *electric current*. Since this is a fundamental quantity, it is not defined any more than we define mass, length, and time. We know from experiment how it will make its presence felt. If we have an electric current flowing near a compass needle the needle is deflected, and if we have currents flowing in two wires there will be a force between them. The electric current, I, appears in the expressions for the dimensions of a physical quantity in the same way as mass, length, and time. For example, the dimension of velocity is length divided by time, written in the notation of dimensional analysis as $[LT^{-1}]$. Similarly the dimension of electric charge is $[TI]$.

The units most commonly used in electrical engineering work today are the meter, kilogram, second, and ampere. Unfortunately, however, the ampere is not usually regarded as a basic unit as the other three are, since there is no recognized standard at present for defining it as an independent physical unit. The ampere is a derived unit based upon the force between two current-carrying conductors. It will be shown in Chapter 6 that the force acting on length l of two infinitely long parallel wires which are separated by a distance r in vacuum and each carrying a current I is given by

$$F = \frac{\mu_0 I^2 l}{2\pi r}.$$

The constant μ_0 is the permeability of vacuum (see Chapter 4) and is conventionally defined as $4\pi \times 10^{-7}$. We would expect to have to assign an arbitrary value to some quantity in defining the unit of a new fundamental quantity. The actual choice made seems curious but results from the historical development of the subject. The unit of current is now defined so as to make the constant μ_0 in the above expression equal to $4\pi \times 10^{-7}$. For our present purposes we may regard this discussion of the standardization of the unit of current as a digression. The important point for us is that we have introduced the electric current as a fourth fundamental quantity, and its unit is the ampere (amp). We may leave the actual manner in which the standard ampere is obtained to the national standards laboratories. In the mksa system the electrical quantities turn out to be the practical units we are already familiar with (volt, ohm, etc.).

We are now in a position to derive the other electrical quantities in terms of the fundamental quantities we have chosen. An electric charge in motion constitutes an *electric current*. This we are accustomed to dealing with in circuit theory. The unit of charge is defined as unit current flowing for unit time, that is, 1 amp-sec, and is called the *coulomb* (coul). If a current flows in a material body, then we know by experiment that heat is produced continuously, and hence power is being expended. We know that if mass flows continuously from a point of higher potential energy to one of lower potential energy, then power is also expended. Hence, by analogy we can conceive of the idea of an *electric potential* and a *potential difference* across the body through which the current flows. Experiment shows the power to be proportional to the current and to the potential difference, and so we have the experimental law

$$P = VI,$$

where P is power, I is current, and V is potential difference. This law is used to define potential difference, and the unit is such that unit power is dissipated when unit current flows. The unit of power is, of course, the *watt* (w) and can be defined in terms of mechanical quantities only. The unit of potential difference is the *volt* (v).

Another law we derive from experiment is that the potential difference across the body is proportional to the current flowing in it. Hence we can write

$$V = RI,$$

where R will be a constant for the body concerned. This expression is the well-known law due to Ohm, which defines the *electrical resistance* of the body. A unit resistance is one which has a unit potential difference (1 v) across it when unit current (1 amp) is flowing through it. The unit of resistance is known as the *ohm*.

The statements of the preceding paragraph need some qualification. The law is only true within limits. If too great a voltage is applied, then breakdown occurs, and the properties of the material are permanently altered. Furthermore, some materials do not obey the law at all. One example is the ceramic material known as thyrite. Materials in which the relation between current and voltage is not linear are said to have a nonlinear resistance. Thus we use the equation above to define an ideal linear resistance. We can then use the equation to deduce the properties of networks containing linear resistances. It should be noted, however, that there would be little point in doing this (except for academic interest) unless it turned out in practice that a large class of materials obeyed the law sufficiently well for the results to be of use to the practicing engineer.

2–3 The electric-field quantities

We saw in the first chapter that we needed to introduce new quantities to deal with the case of current fields. We described the direction of the current flow by using a vector. We also introduced the *current density*, which is the current per unit area and is readily defined in terms of the fundamental units. The unit of current density will therefore be the amp/m^2. We also introduced the **E**-vector, and its unit is the v/m.

By using Ohm's law we deduced the relation between **E** and **J**, namely, $\mathbf{J} = \sigma\mathbf{E}$. It follows from this equation that the unit of conductivity (σ) is $(\text{ohm-m})^{-1}$. Let us consider a cube of material as shown in Fig. 2–1, and let each side of the cube be 1 m. Suppose that a current is introduced into the material via a perfectly conducting plate placed over one face, and leaves via a second plate over the opposite face. If the material is homogeneous; i.e., the conductivity is the same throughout the cube, then we can assume the fields to be uniform. Thus the magnitude of the electric field is V, the length of the cube being unity, and the magnitude of the current density is I, the cross section also being unity. Hence $\sigma = 1/R$. We can think of the conductivity as being equal to the conductance of a homogeneous cube whose edge is 1 m long, the current entering and leaving uniformly across one pair of faces.

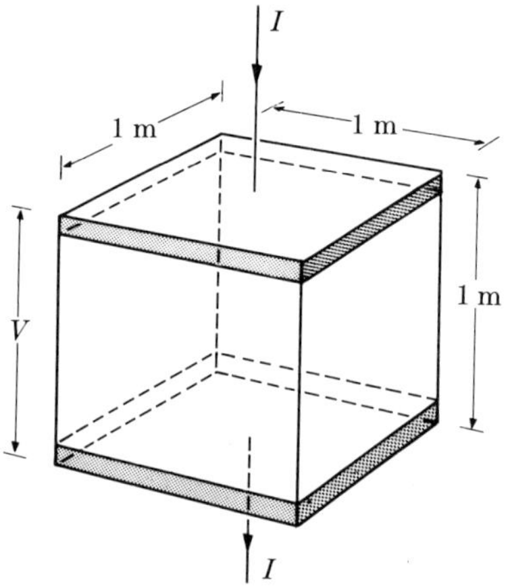

FIG. 2–1. A 1-m resistive cube.

So far we have supposed that the vectors **E** and **J** are in the same direction. This is not necessarily always the case. The current may be constrained in some way, perhaps by the properties of the medium, so that it flows in a different direction from the force vector **E**. To calculate the power in this case, we consider a cylinder of infinitesimal dimensions with axis parallel to the **J**-vector. The current flowing in this cylinder will be $J\,dS$. The voltage across it will be $E \cos\theta\, dl$, where $\cos\theta$ is the angle between the **E**- and **J**-vectors. Hence the power will be $EJ\cos\theta\, dS\, dl$. Since the volume power density p is given by

$$p = \lim_{\Delta\tau\to 0} \frac{P}{\Delta\tau},$$

and $dS\,dl$ is the volume, we have

$$p = \mathbf{E}\cdot\mathbf{J}.$$

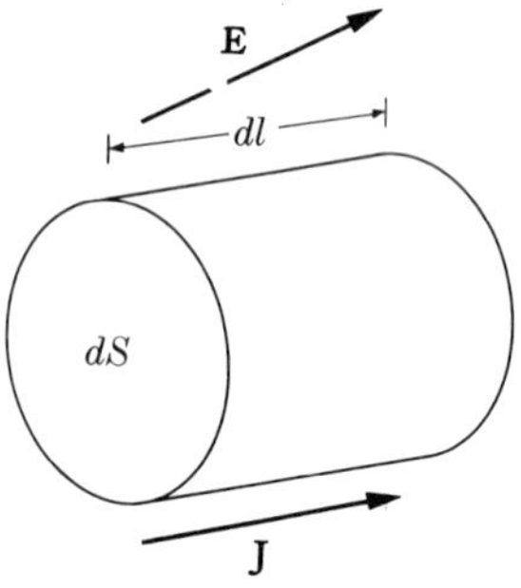

FIG. 2–2. The general case where the electric field and the current density are in different directions.

The relations for circuits and fields are compared in Table 2–1.

TABLE 2–1

Circuit	Fields
$I = GV$	$\mathbf{J} = \sigma\mathbf{E}$
$P = VI$	$p = \mathbf{E} \cdot \mathbf{J}$

We have now developed our ideas of field theory to a point corresponding to the one reached in circuit theory when all the circuit elements are resistive.

Problems

2–1 Derive the dimensions of (a) power, (b) potential difference, (c) resistance, and (d) electric charge, in terms of the basic quantities mass, length, time, and current.

2–2 Derive the dimensions of (a) current density, (b) electric field, (c) conductivity, and (d) resistivity, in terms of the basic quantities.

CHAPTER 3

The Capacitive Properties of the Field

3–1 Introduction

In the preceding chapters we developed the ideas of fields in terms of the flow of electric charge or the conduction current. The basic experimental law used was Ohm's law, and hence we might regard the behavior of the fields as being governed by the resistive properties of the medium. In this chapter we shall see that fields may exhibit what we could regard as capacitive properties governed by the capacitive properties of the medium.

3–2 The capacitor and the corresponding fields

The basic law for capacitors corresponding to Ohm's law for resistors is $Q = CV$. This too is an experimental law which states that the voltage across a capacitor is proportional to the charge on the plates. The law is not necessarily obeyed for all capacitors in practice, but we can use the equation first to define an ideal linear capacitor and then to obtain the behavior of circuits containing ideal capacitors. The practical utility of the results arises because a large class of capacitors does obey the law to a very good approximation under actual working conditions. The equation defines the capacitance C of the condenser, and the unit of capacitance is the capacitance of a condenser which has unit potential difference across its plates when there is unit charge on them. The unit is called the *farad* (f) after Faraday, and we now have it defined in terms of our fundamental quantities.

We know from experiment that work must be done in order to charge the capacitor, and therefore the capacitor stores energy. We can calculate this as follows. If a current I is flowing into a capacitor with a voltage V across it, then the source is supplying power VI. Hence in time dt the energy supplied by the source is $VI\ dt$, and if we suppose that there are no resistive elements in the system so that no energy is dissipated, then this energy must be stored in the capacitor. But $I = dQ/dt$ by definition, and thus the energy dW is given by $V\,dQ$. We also have the experimental law relating V and Q, and hence

$$dW = \frac{Q}{C}\,dQ.$$

If the initial charge on the capacitor is zero, the final charge is Q, and the final voltage is V, then

$$W = \int_0^Q \frac{Q}{C}\,dQ = \frac{1}{2}\frac{Q^2}{C} = \frac{1}{2}CV^2.$$

To obtain the corresponding fields we consider a rectangular capacitor as shown in Fig. 3-1. Again we assume idealized conditions and hence that the charge is distributed uniformly over the plates and the potential drop across the intervening medium is uniform. We thus get a vector $\mathbf{E}$ of magnitude V/l. Hence, from the basic relation for the capacitor $Q = CV$, we have $Q = ClE$.

In analogy with our treatment of the resistive case, we divide this expression by the area of the plates, obtaining

$$\frac{Q}{A} = \frac{Cl}{A}E.$$

We now define two new quantities, again in an analogous fashion:

$$D = \frac{Q}{A} \quad \text{and} \quad \epsilon = \frac{Cl}{A}.$$

Our equation then becomes

$$D = \epsilon E,$$

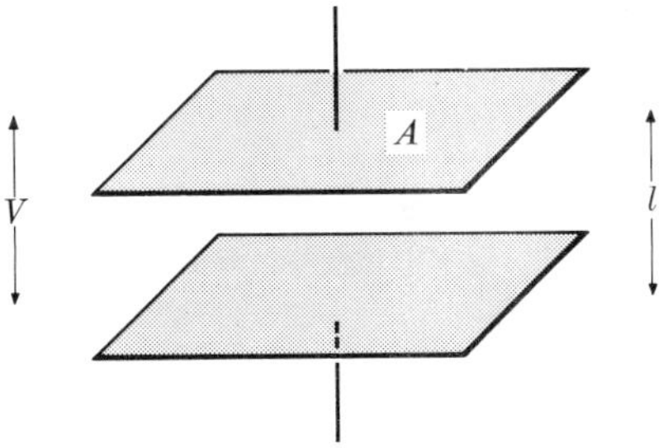

FIG. 3-1. The parallel plate capacitor.

and we have D analogous to J and ϵ to σ. The quantity E appears in the equation for resistors. This is the electric-field vector which occurs in electrostatics. The quantity ϵ is known as the *permittivity* and, like conductance, is a property of the medium. It is equal to the capacitance of a capacitor whose plates have an area of 1 m^2 and are placed 1 m apart. The D-vector is the counterpart of the J-vector and may be considered to be a flow vector. It is called the *electric flux (flow) density*. The formal similarity of the electrostatic equation $D = \epsilon E$ and the electric current equation $J = \sigma E$, and their relation to the phenomena of hydrodynamics were clearly seen by the early workers who used these facts extensively. The reason for the use of the term and the idea of flux in electrostatics lies entirely in this analogy. There is nothing physically flowing or moving, and it is understandable that we find our first encounter with an electrostatic (stationary) flux (flow) a little confusing. The equations above also define the units in terms of quantities which either are themselves fundamental quantities or have been defined in terms of fundamental quantities. Thus the unit of permittivity ϵ is f/m, and of flux is coul/m^2. We have defined ϵ and D in only a crude way so far. Refinement will be needed in order to define ϵ and D at a point. This will enable us to define point functions upon which we based our ideas of fields. However, before doing this, we shall first relate the electric field E as introduced here to the definition used in classical electrostatics.

3-3 The electrostatic field

We know from mechanics that if we move a force F a distance $-dl$, that is, in the opposite direction to that in which the force is acting, then we do work on the force, and the energy in the system increases, and therefore $dW = -F\,dl$. Hence from the last section we have

$$F = -\frac{dW}{dl} = -\frac{d}{dl}(\tfrac{1}{2}CV^2) = -CV\frac{dV}{dl} = QE.$$

This means that if the condenser is charged by taking electric charge from one plate across the intervening medium to the other plate, then there will be a force acting on this charge equal to QE throughout the process of transfer. Hence we see that the electric field may be defined as the force acting per unit charge. If we wanted to explore the field between the plates by introducing a test charge from outside into the region, then we would make the refinement that only an infinitesimal charge be introduced so that the existing field is not disturbed by the presence of the test charge. Hence we may define the electric field at a point as the force acting per unit charge when an infinitesimal charge is placed at that point. We thus reproduce the classical definition of the electric field.

3-4 Energy in the field

Consider again the capacitor of Section 3-2. We have shown that the energy stored in the capacitor is $\frac{1}{2}QV$. In terms of the field quantities we can write this as

$$W = \tfrac{1}{2}QV = \tfrac{1}{2}DAEl.$$

But Al is the volume of the capacitor, and the energy per unit volume or energy density is therefore given by

$$w = \tfrac{1}{2}DE = \tfrac{1}{2}\epsilon E^2.$$

We have not proved that the stored energy is in fact uniformly distributed throughout the volume, and no direct experimental test can be made. However, the idea of an energy density at a point given by this formula has proved useful in practice, and it may be considered verified by the results to which it leads.

3-5 Permittivity and flux density

We saw that by considering an ideal capacitor we could introduce quantities ϵ and D formally similar to the conductivity σ and the current J which we obtained in considering an ideal resistor. To obtain quantities which are useful in field theory we now need to take the limit as the dimensions of the capacitor become infinitesimally small. We saw that ϵ could be regarded as the capacity of a cubic condenser with side 1 m in length if we suppose the medium to be homogeneous, that is, ϵ is the same throughout the medium. If the cube had a side Δl then the capacity ΔC would be related to the permittivity ϵ by the equation

$$\epsilon = \frac{\Delta C \cdot \Delta l}{(\Delta l)^2} = \frac{\Delta C}{\Delta l},$$

and hence, to obtain the value of ϵ at a point, we would take the limit

$$\epsilon = \lim_{\Delta l \to 0} \frac{\Delta C}{\Delta l}.$$

We have now defined both ϵ and E at a point, and can define the magnitude of D at that point by the relation

$$D = \epsilon E.$$

The quantity D was formally related to the flow vector, and hence if D is to be a vector, we must also define its direction. For the present we will suppose that its direction is the same as the direction of $\mathbf{E}$. However, just as with $\mathbf{J}$ this is not necessarily always true, but it is true for a large class of media occurring in practice. These media are known as *isotropic media*. We then have the vector equation

$$\mathbf{D} = \epsilon \mathbf{E}.$$

3–6 Gauss' theorem and Coulomb's law

The unit of $\mathbf{D}$ is charge per unit area, and thus we might suspect that $\mathbf{D}$ is associated in some way with the presence of charge in the system ($\mathbf{J}$ being associated with a time rate of change of charge). Let us again consider the capacitor of Fig. 3–1. Suppose that we enclose one of the plates with any closed surface, the only restriction on this surface being that we specify that between the two plates it shall be a plane surface parallel to the plates. The flux density $\mathbf{D}$ is zero everywhere over this surface except on the plane part between the plates. Here it is perpendicular to the surface, and therefore the total outward flow over the surface is DA. But from Section 3–2 this is equal to Q, the charge on the plate. Since $\mathbf{D}$ behaves like a flow vector, it follows that if the surface between the plates were arbitrary, we would get the same result by calculating $\iint \mathbf{D} \cdot d\mathbf{S}$. Thus we have the result that the integral of the flux density over a closed surface is equal to the charge enclosed by the surface:

$$\oint\!\!\!\oint \mathbf{D} \cdot d\mathbf{S} = Q.$$

This is Gauss' law and is of general applicability. Note that if the flow is outward, it implies that a positive charge is enclosed. It is also of interest to note that in the above argument, if we had enclosed the other plate of the condenser, then the sign of the flow would change. But we know that the other plate of the condenser carries an opposite charge, and so the result is consistent.

We shall now derive the differential form of Gauss' theorem. Let us assume that we apply this theorem to a small volume element $\Delta\tau$, and suppose that within this element there is a charge density q_v, the element being small enough for this to be considered a constant. The charge enclosed is then $q_v\,\Delta\tau$. Thus by Gauss' theorem,

$$\oiint \mathbf{D}\cdot d\mathbf{S} = q_v\,\Delta\tau.$$

Now, taking the limit as $\Delta\tau \to 0$, we have

$$\lim_{\Delta\tau\to 0}\frac{\oiint \mathbf{D}\cdot d\mathbf{S}}{\Delta\tau} = q_v,$$

and so

$$\operatorname{div}\mathbf{D} = q_v.$$

This equation should be compared with the corresponding relation for $\mathbf{J}$, namely, $\operatorname{div}\mathbf{J} = -\partial q_v/\partial t$. We see that q_v constitutes the source of the $\mathbf{D}$-field (see Chapter 1).

Let us next consider a point charge Q. From the symmetry of this case we see that the magnitude of D will depend only upon the distance from the charge, and that it will be directed radially. Let us consider a spherical surface of radius r centered on the charge, and apply Gauss' theorem:

$$\oiint \mathbf{D}\cdot d\mathbf{S} = 4\pi r^2 D = Q,$$

$$\therefore D = \frac{Q}{4\pi r^2},$$

and since $D = \epsilon E$, we have the electric field

$$E = \frac{Q}{4\pi\epsilon r^2}.$$

If a second charge Q' is placed distant r from the first one, then it experiences a force F given by $Q'E$, or

$$F = \frac{QQ'}{4\pi\epsilon r^2}.$$

This is the law due to Coulomb, which is of great historical importance. Coulomb verified by direct experiment that the force between two charges was directly proportional to the magnitude of the charges and inversely proportional to the square of the distance between them. This formed the original starting point for the science of electrostatics. Here we began from different experimental facts and deduced this law. As noted earlier, there is no one set of laws or postulates from which one must start, but wherever one starts the results predicted by the theory must always check with the experimental results any time a result capable of direct verification is obtained. The fact that we deduced

Coulomb's law, a law which has already been shown to be in agreement with experiment, is a verification of our original postulates or experiments.

Coulomb's law, as stated above, does not give the complete equation but shows only that

$$F \propto \frac{QQ'}{r^2}.$$

Further experiments demonstrated that this force was modified by the medium in which the charges were situated, and a dimensionless factor κ, known as the *dielectric constant*, was introduced into the denominator to account for this. The dielectric constant of a material medium is found by experiment to be always greater than that of a vacuum. As far as Coulomb's experiments went there was no need for any further complication of the formula, and the constant of proportionality was taken to be dimensionless and to have magnitude unity. This, however, did not prove to be convenient when the later work on the relation between electric and magnetic fields was done. By starting with the mksa system of units and following the presentation given here, permittivity automatically turns out to be a quantity with dimensions, and the factor of 4π (originally introduced to "rationalize the units") arises naturally. In this system, a vacuum has a permittivity ϵ_0, which is determined experimentally to be 8.85×10^{-12} f/m. The dielectric constant is then defined as the ratio of the permittivity of the medium to that of vacuum, and so retains its dimensionless character:

$$\kappa = \frac{\epsilon}{\epsilon_0}.$$

3–7 Boundary conditions

The boundary condition for the normal component of the flux density is readily obtained by applying Gauss' theorem. Let us consider the boundary between two media as shown in Fig. 3–2. Now we construct a cylindrical surface across the boundary with cross section ΔS, axis perpendicular to the boundary, and length Δl, and we apply Gauss' theorem to this surface. The contribution to the surface integral from the end of the cylinder in Region 1 will be $D_1^n \, \Delta S$, where D_1^n is the normal component of **D** in Region 1. The contribution from the end in Region 2 will be $-D_2^n \, \Delta S$, the negative sign arising because if the two vectors are pointing in the same direction, D_1^n will give a flux

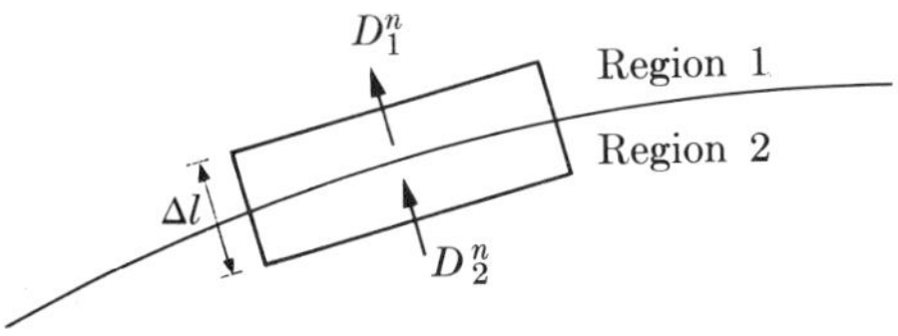

FIG. 3–2. The boundary condition for **D**.

out of the volume element, and D_2^n a flux into it. In order to get the difference between D_1^n and D_2^n at the boundary, we must let $\Delta l \to 0$, so that D_1^n is at the boundary in Region 1 and D_2^n at the boundary in Region 2. The contribution to the integral from the curved surface of the cylinder then vanishes, and the integral becomes $(D_1^n - D_2^n)\,\Delta S$. By Gauss' theorem this integral is equal to the charge enclosed by the surface, which will be finite in the limit $\Delta l \to 0$ if there is a surface charge density at the boundary. If there is a charge q_s per unit area contained in a sheet of zero thickness coincident with the boundary, then the charge enclosed by the surface is $q_s\,\Delta S$. Hence

$$(D_1^n - D_2^n)\,\Delta S = q_s\,\Delta S, \qquad \text{or} \qquad D_1^n - D_2^n = q_s.$$

There is then a discontinuity in the normal component of the electric flux at a boundary equal to the surface charge density at the boundary.

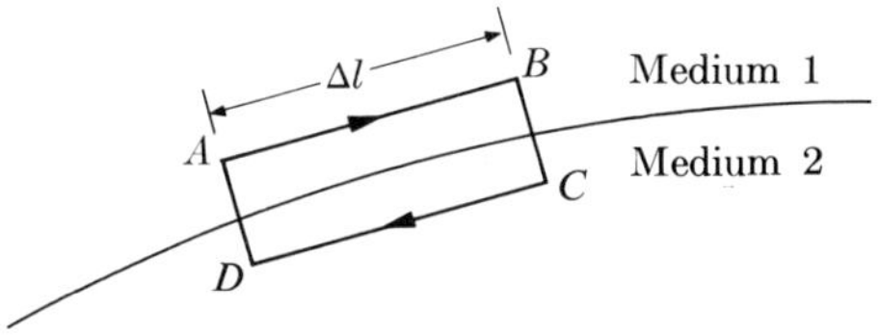

FIG. 3–3. The boundary condition for **E**.

The boundary condition for **E** in the electrostatic case is obtained by applying the relation $\oint \mathbf{E} \cdot d\mathbf{l} = 0$. We take the integral around the contour $ABCD$, as shown in Fig. 3–3. If we suppose that the length vector is in the direction AB, then the contribution to the integral from the path AB is $E_1^t\,\Delta l$, where E_1^t is the tangential component (i.e., tangential to the boundary surface) in Medium 1. Along CD, $d\mathbf{l}$ will be negative, and the contribution will therefore be $-E_2^t\,\Delta l$. In order to get the values of E_1^t and E_2^t at the boundary we must make BC and DA vanishingly small, and so the contribution to the integral from these two sides is zero. Hence

$$E_1^t\,\Delta l - E_2^t\,\Delta l = 0, \qquad \text{or} \qquad E_1^t = E_2^t.$$

The tangential component of **E** is continuous at the boundary. This same result also holds true for time-varying fields, but the proof must be left until later.

Problems

3–1 Derive the dimensions of electric flux density and permittivity in terms of the basic quantities.

3–2 Use Gauss' theorem to calculate the field at all points due to a uniform charge density within a sphere of radius a.

3–3 What is the value of the point charge at the origin which will make the field of Problem 3–2 zero at all points outside the sphere? Sketch the curve showing the variation of the field inside the sphere as a function of distance from the center.

3–4 Is the field of Problem 3–2 (a) solenoidal, (b) irrotational?

3–5 The charge density in Problem 3–2 is 1 coul/m^3 and the sphere has radius 1 m. Calculate the field at the following distances from the center of the sphere: (a) 0.5 m (b) 1 m, (c) 10 m.

3–6 An infinite plane carries a uniform surface charge density q_s per unit area. Using Gauss' theorem, calculate the electric field.

3–7 Calculate the potential in Problem 3–6.

3–8 Given that the charge density in Problem 3–6 is 1 coul/m^2, calculate the potential at 1 m and 10 m from the plane.

3–9 Calculate the energy stored in the field of Problem 3–2 as a function of position. What is the total energy stored within the sphere of radius a?

3–10 Calculate the total energy stored within the sphere for the values given in Problem 3–5.

3–11 Calculate the field due to an infinite straight line charge of density q_l per unit length.

3–12 Use Coulomb's law to determine the field on the line joining two equal and opposite charges.

3–13 An electric field is given by $k_1\mathbf{u}_x$ for $x < 0$ and $k_2\mathbf{u}_x$ for $x > 0$, and the permittivity is ϵ_1 for $x < 0$ and ϵ_2 for $x > 0$. Calculate the charge density. What is the condition for zero charge density?

3–14 Calculate the force on two charges of 1 coul separated by 1 m in vacuum. Give the result in newtons and in kilogram-weight.

CHAPTER 4

The Inductive Properties of the Field

4–1 Introduction

We have seen that we can develop fields governed by the resistive and capacitive properties of the medium by extending the circuit laws for the resistor and the capacitor. The third type of element we deal with in circuit theory is the inductor, and in this chapter we develop fields controlled by the inductive properties of the medium by extending the circuit law for the inductor.

4–2 The inductor and the corresponding fields

Again we start from the experimental law governing the relation between the current in an inductor and the voltage across it, that is, $V = L(dI/dt)$. In exactly the same way as with resistors and capacitors, this law is not necessarily obeyed under all conditions but can be used to define an ideal linear inductor. Since the law is obeyed well enough in practice for a large class of inductors, its study is useful. This equation then defines the inductance L and also the unit of inductance, which is such that a unit voltage across the inductor produces a unit rate of change of current flowing in it, that is, 1 v causes the current to change at the rate of 1 amp/sec. The unit is called the *henry* (h), and we now have it defined in terms of fundamental quantities.

We know from experiment that work must be done to build up a current in an inductor, and hence the inductor stores energy. We can calculate this work as follows. If a current I is flowing into an inductor with a voltage V across it, then the source is supplying power VI. In time dt therefore the energy supplied by the source is $VI\,dt$, and if we suppose no resistive elements in the system so that no energy is dissipated, then the energy must be stored in the inductor. Hence, since $V = L(dI/dt)$,

$$W = \int VI\,dt = \int LI\frac{dI}{dt}\,dt = L\int_0^I I\,dI,$$

$$\therefore W = \tfrac{1}{2}LI^2.$$

We may compare this with the result $W = \frac{1}{2}CV^2$ for capacitors.

For the capacitor, we were able to introduce the new field quantities by analogy with the resistive case. We shall now introduce fields for the inductive case by analogy with the capacitive case, bearing in mind the dual nature of the two cases. The formulas will be listed in pairs to emphasize the analogy. We first compare the basic equations:

$$I = C\frac{dV}{dt}, \qquad V = L\frac{dI}{dt}.$$

Since we found it convenient to work with the capacitor in terms of the parameter Q, we introduce the corresponding quantity for the inductor,

$$Q = \int I\,dt, \qquad \psi = \int V\,dt,$$

and the basic equations then become

$$Q = CV, \qquad \psi = LI.$$

We now arrange the capacitor equation as we did in the last chapter, and make the corresponding arrangement of the one for the inductor:

$$\frac{Q}{A} = \frac{Cl}{A}\frac{V}{l}, \qquad \frac{\psi}{A} = \frac{Ll}{A}\frac{I}{l}.$$

The field quantities are defined as before:

$$\frac{V}{l} = E, \qquad \frac{I}{l} = H,$$

$$\frac{Q}{A} = D, \qquad \frac{\psi}{A} = B,$$

$$\frac{Cl}{A} = \epsilon, \qquad \frac{Ll}{A} = \mu.$$

The above equations can now be written

$$D = \epsilon E, \qquad B = \mu H.$$

We have not yet seen how to associate a direction with B or H, but if we anticipate that they will be vectors, then we have obtained a new flux vector **B**, a new force vector **H**, and a new scalar quantity μ. These are known as the magnetic flux density (**B**), the magnetic field (**H**), and the permeability (μ). The term *magnetic* arises as a result of the historical development of the subject. We know from experiment that a magnetic compass needle placed in the region near an inductor will be deflected.

The units of the new quantities are readily obtained from the definitions above. The unit of H is the amp/m, that of B is the v-sec/m^2 or the weber/m^2, and that of μ the h/m. Our definitions are as yet crude, and we must refine them as we did for D and ϵ in the last chapter.

4–3 Energy in the field

We proceed in a manner analogous to the treatment given in Section 3–4. From Section 4–2, for the energy in the inductance, we have

$$W = \tfrac{1}{2}LI^2 = \tfrac{1}{2}\psi I.$$

Writing this in terms of the field quantities, we have

$$W = \tfrac{1}{2}BAHl.$$

But Al is volume, and so the energy per unit volume or the energy density is given by

$$w = \tfrac{1}{2}BH = \tfrac{1}{2}\mu H^2.$$

4–4 Permeability and flux density

We refined our definitions of permittivity and electrostatic flux density by considering again the capacitor and then allowing its dimensions to become infinitesimal. We now carry out the same procedure for the magnetic quantities. Just as we considered the simplest type of capacitor consisting of only two plates, so we shall consider the simplest type of inductor consisting of a single turn (Fig. 4–1). We can measure the inductance by observing V and I and using the relation $V = L(dI/dt)$. From Section 4–2 the relation between the permeability and the inductance is given by

$$\mu = \frac{\Delta L\,\Delta l}{\Delta A},$$

and if we construct our inductance so that $\Delta A = (\Delta l)^2$, then

$$\mu = \frac{\Delta L}{\Delta l},$$

and we may define the permeability at a point by the relation

$$\mu = \lim_{\Delta l \to 0} \frac{\Delta L}{\Delta l}.$$

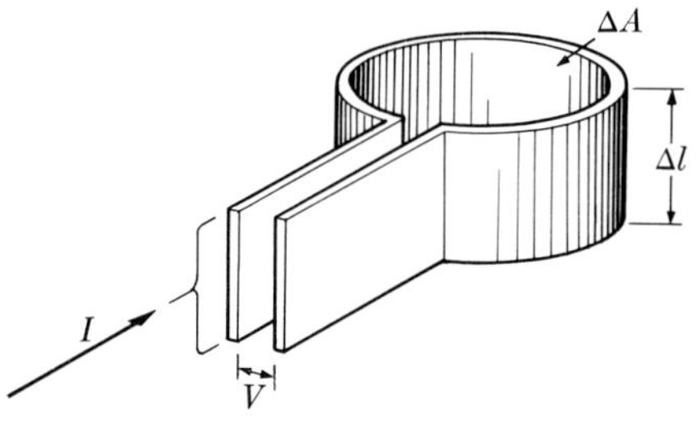

FIG. 4–1. The single-turn inductor.

We can also observe $\int V\,dt$ and hence obtain ψ. The magnetic flux density B is then given by $\psi/\Delta A$, and so in the limit as the dimensions of the coil become infinitesimal, we can define B at a point. Finally, from the relation $B = \mu H$, we can define H at a point.

So far we have treated B and H as having only magnitudes. The inductance illustrated in Fig. 4–1 is, however, associated with a direction which can be defined by the axis of the cylinder. We can obtain a vector $\mathbf{B}$ by associating this direction with it, the sense being obtained from the right-hand rule. For the present we will suppose that the direction of $\mathbf{H}$ is the same as that of $\mathbf{B}$.

Again this is not universally true, but it defines a magnetically isotropic medium, a property possessed by a large number of materials in practice. Thus we have the vector equation

$$\mathbf{B} = \mu \mathbf{H}.$$

4–5 Origin of the field

In considering the electric field we saw that the source of the field was the charge density, and we could consider the flux lines to begin and end on a charge. In the previous section we produced a magnetic field by means of an electric current. It would therefore appear that the electric current is the "origin" of the field in some way, but the flux lines pass through the loop and are left "hanging." There is no way of terminating them on the current. To overcome this difficulty let us take a number of inductors of the type shown in Fig. 4–1 and place them end to end, the last one meeting the first so as to form a closed surface. (The figure produced would be a toroid or a doughnut.) The H- and B-lines will now form closed curves so that we have avoided the problem of where to terminate them. For the ith inductor we would have from our definition

$$H_i = \frac{\Delta I_i}{\Delta l_i},$$

and summing over all the inductors, we obtain

$$\sum_i H_i \Delta l_i = \sum_i \Delta I_i.$$

In the limit as $\Delta l \to 0$, we have

$$\oint H\, dl = I,$$

where the integral is now taken around a closed curve, and I is the total current enclosed by that curve. We recognize this as a special case of Ampère's law. Again we have produced a result which has been verified experimentally and so have another verification of our theory. The more general relation obtained experimentally by Ampère states that

$$\oint \mathbf{H} \cdot d\mathbf{l} = I.$$

The left-hand side is a line integral taken around any closed contour in the $\mathbf{H}$-field, and $\mathbf{I}$ is the total current enclosed by the contour. If we have a general current field represented by $\mathbf{J}$, then the current enclosed by the contour is given by $\iint \mathbf{J} \cdot d\mathbf{S}$ taken over a surface bounded by the contour. If we consider a contour of very small dimensions enclosing an area ΔS, we have

$$\oint \mathbf{H} \cdot d\mathbf{l} = \mathbf{J} \cdot \mathbf{\Delta S}.$$

We can make this a maximum by orientating the contour so that $\mathbf{J}$ and $\Delta\mathbf{S}$ are in the same direction, thereby maximizing the scalar product. Dividing by ΔS and taking the limit as $\Delta S \to 0$, we have

$$\lim_{\Delta S \to 0} \frac{(\oint \mathbf{H} \cdot d\mathbf{l})_{\max}}{\Delta S} = \mathbf{J},$$

and so

$$\operatorname{curl} \mathbf{H} = \mathbf{J}.$$

It is interesting to note at this point that we related a contour integral to a surface integral bounded by the contour. However, the surface remains undefined except at its boundary. We saw earlier that such a surface integral depends only on the boundary of the surface, only if the vector integrated has zero divergence. Hence we might expect our result to apply only if div $\mathbf{J} = 0$, and since div $\mathbf{J} = -\partial q_v/\partial t$, div $\mathbf{J}$ will be equal to 0 if $\partial/\partial t = 0$, that is, for static fields. This is in fact the case. The above equation is incomplete for time-dependent fields, as was shown by Maxwell. We shall return to this point in a later chapter.

The **H**-field is an example of the type of field mentioned in Section 1-6. The current density $\mathbf{J}$ is the origin of the **H**-field, and $\mathbf{J}$ is referred to as the vortex of the **H**-field. We may contrast this with the relation div $\mathbf{D} = q_v$, where q_v is referred to as the source of the **D**-field. Thus we have two types of field, one arising from sources and one from vortexes. (The general field would have both sources and vortexes.)

4-6 Boundary condition for H

We can apply Ampère's law to determine the boundary condition for **H**. Let us consider the boundary between two media as shown in Fig. 4-2. Suppose that there is a current flowing at the surface separating the two media; by this we mean that the current flows in a sheet of zero thickness. The current density $\mathbf{J}$ will therefore be infinite because we are supposing a finite current in an infinitely thin sheet. Hence we define a surface current density $\mathbf{J}_s$, which is the current per unit length crossing a line drawn in the sheet perpendicular to the current. To obtain the surface current density at a point we would take a limit. Now we construct a contour $ABCD$ perpendicular to the current, and apply Ampère's law. To determine the direction of integration around the contour we use the right-hand rule, and thus if the current flows down into the page, the direction of integration will be as shown. If we suppose that **H**-vectors pointing to the right are positive, then for the integration along AB, $\mathbf{H}$ and $d\mathbf{l}$ are in the same direction, and along CD they are opposite. Let $H_\perp^{(1)}$ be the component of **H** in Medium 1, perpendicular to the current, i.e., in the AB-direction. Then the contribution to the integral from the side AB is $H_\perp^{(1)}\,\Delta l$, where Δl is the length of AB. Similarly for CD the contribution is $-H_\perp^{(2)}\,\Delta l$. We can shrink AD and BC down to zero, and the rectangle will still enclose a finite current, since the current flows in a sheet of zero thickness at the surface. Hence the contributions to the integral from these two sides is made zero. The

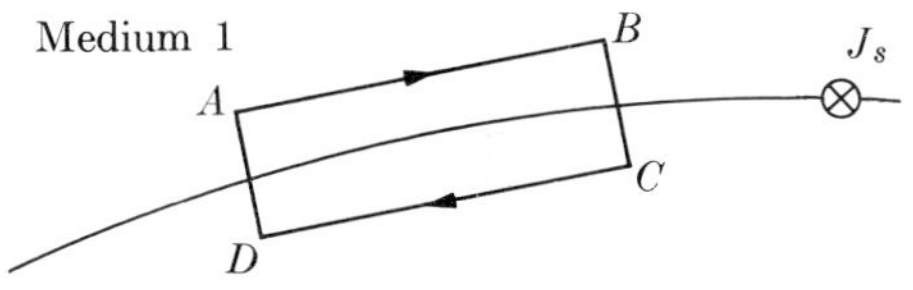

FIG. 4–2. The boundary condition for **H**.

current enclosed by the contour is $J_s\,\Delta l$, since we constructed the contour perpendicular to the current. Thus

$$(H_\perp^{(1)} - H_\perp^{(2)})\,\Delta l = J_s\,\Delta l, \qquad \text{and} \qquad H_\perp^{(1)} - H_\perp^{(2)} = J_s.$$

If the rectangle is taken parallel to the current, we have $H_{||}$, and clearly there is no current flowing through the contour; hence

$$H_{||}^{(1)} = H_{||}^{(2)}.$$

Thus on passing through a current sheet there is a discontinuity in the tangential component of **H** perpendicular to the direction of the current and of a magnitude equal to the surface current density $\mathbf{J}_s$.

4–7 Flux linkage and mutual inductance

We saw above that the energy associated with an inductance L carrying a current I is $\frac{1}{2}LI^2$. Using our elementary inductance (Fig. 4–1) we developed the idea of a flux crossing the area enclosed by the current. In Chapter 1 we saw that a field could be represented graphically by flux lines whose density is proportional to the magnitude of the flux density. Thus we can imagine lines representing the magnetic flux passing through the circuit around which the current flows. Such a flux line is said to be linked with the circuit, and so we get the idea of flux linkage. The flux is given by $\psi = LI$, and hence a circuit carrying a current I has a flux LI linked with it. From the relation given above for energy the work done in establishing a flux linkage ψ with a circuit carrying a current I is $\frac{1}{2}\psi I$.

So far we have only considered the flux linked with the current producing it. The inductance L which occurs in the expressions above is accordingly sometimes referred to as the *self-inductance*. Since we can produce with one current a magnetic flux linked with a different current, we are led to the concept of a *mutual inductance*, a quantity with which we are familiar from circuit theory. If we have two separate circuits, then the equations for the voltages and currents can be written

$$V_1 = L_1\frac{dI_1}{dt} + M_{12}\frac{dI_2}{dt}, \qquad V_2 = L_2\frac{dI_2}{dt} + M_{21}\frac{dI_1}{dt}.$$

Suppose we start with the currents in both circuits equal to zero. We first increase the current in circuit 1 to the value I_1, with circuit 2 an open circuit.

The work done by generator 1 is given by

$$\int V_1 I_1\, dt = \int L I_1\, dI_1 = \tfrac{1}{2} L I_1^2.$$

No work is done by generator 2 since it supplies no current. We then increase the current in circuit 2 to the value I_2, keeping I_1 constant. The work done by generator 1 is given by

$$\int V_1 I_1\, dt = \int M_{12} I_1\, dI_2 = M_{12} I_1 I_2,$$

and that by generator 2 is

$$\int V_2 I_2\, dt = \int L_2 I_2\, dI_2 = \tfrac{1}{2} L I_2^2.$$

Hence the total work done, and therefore the total energy in the system in the final state, is given by

$$W = \tfrac{1}{2} L_1 I_1^2 + \tfrac{1}{2} L_2 I_2^2 + M_{12} I_1 I_2.$$

If we had increased the current in circuit 2 to I_2 first and then that in circuit 1 to I_1, we would have obtained

$$W = \tfrac{1}{2} L_1 I_1^2 + \tfrac{1}{2} L_2 I_2^2 + M_{21} I_1 I_2.$$

But since the final state of the circuit is the same in each case, the two must be equal, from which we have

$$M_{12} = M_{21}.$$

Now we can regard MI_1 as ψ_{12}, the flux due to I_1 linked with I_2, and MI_2 as ψ_{21}, the flux due to I_2 linked with I_1. We see then that work must be done in establishing a flux linkage with a current whether the flux is due to that current or another. Note that M can be either positive or negative depending on whether or not the flux and the current are related by the right-hand rule. Consider, for example, the case shown in Fig. 4-3. If the current I_2 is caused by the linking of B_1 with the circuit, then we can show that I_2 must flow in the direction shown. If it flows in the reverse direction, then it would cause a flux B_2 in the direction of B_1. This, by the same supposition, would cause I_1 to increase, which would make B_1 increase, and this in turn causes I_2 to increase. Such a situation would be unstable, which we know is not the case. Thus the flux due to the induced current is in such a direction as to oppose the change of flux causing the current; i.e., if B_1 increases, it produces a current whose magnetic flux is in the direction opposite to B_1 and so tends to prevent the increase. This result is known as Lenz's law after its discoverer.

Let us again consider the system of Fig. 4-3. When the current in circuit 1 was brought to the value I_1, we saw that a current I_2 was produced in circuit 2 in the opposite direction. If we put a resistor into circuit 2, the current I_2 will

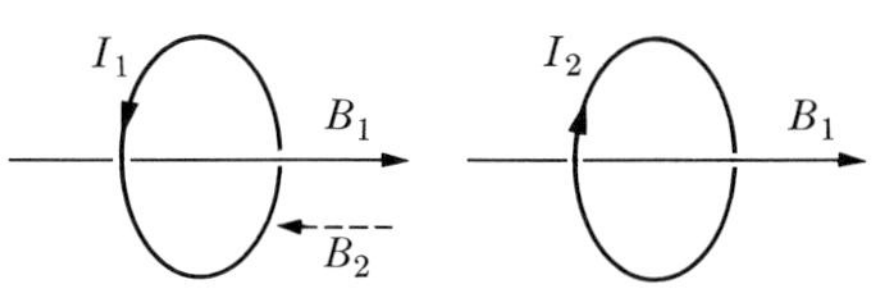

FIG. 4-3. Lenz's law.

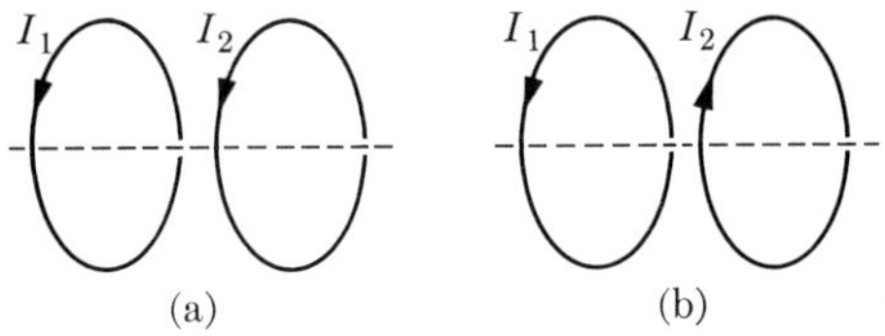

FIG. 4-4. Stable (a) and unstable (b) current configuration.

decay to zero as energy is dissipated in the resistor. The energy turned into heat represents energy removed from the field system associated with the currents; i.e., reducing the current I_2 to zero results in a field system of lower energy. We can look at this result from a different viewpoint. Suppose that both circuits contain batteries which can be adjusted to give the currents desired and which are connected so that both produce currents in the direction of I_1. If we adjust the second circuit so that the current in it remains zero, then the current I_2 which is produced, as discussed above, must flow back into the battery in circuit 2; i.e., it charges the battery. This represents energy put back into the battery, and so the resulting field system has correspondingly less energy. If we bring the current in the second circuit up to the value I_2, we produce the current system shown in Fig. 4-4(a). This, in turn, produces a current in the opposite direction to I_1, and charges the battery in circuit 1, and this represents more energy removed from the field system. If, however, the battery in circuit 2 is reversed in polarity, then the induced current due to producing I_1 in circuit 1 is in the discharge direction for the battery in circuit 2, and so will represent energy flowing out of the battery into the field. A similar result is obtained for the current induced in circuit 1 by producing I_2 in circuit 2. The final current configuration is shown in Fig. 4-4(b). The result of this argument is that owing to the mutual inductance terms, the potential energy of the configuration of Fig. 4-4(a) is less than that of Fig. 4-4(b). We know from mechanics that Fig. 4-4(a) represents the stable equilibrium. Alternatively, we can say that two current-carrying coils suspended so that they can rotate will tend to align themselves so that the planes of the coils are parallel, with the currents in the coils flowing in the same direction. This is obviously readily checked by experiment which shows it to be true. We can also obtain the direction of the forces on the currents. This is done most simply by putting the coils in T-formation as shown in Fig. 4-5, which represents a section of two coils

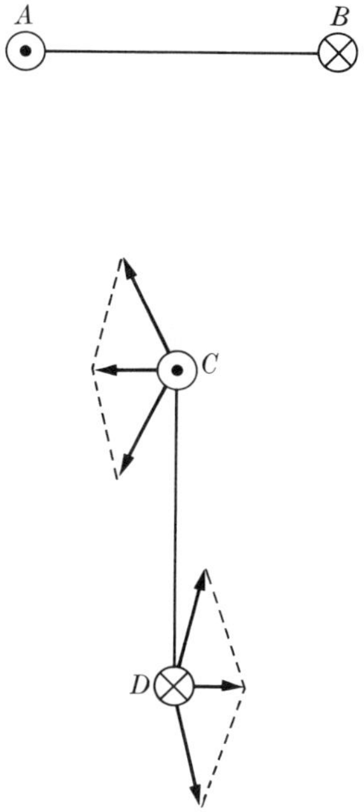

FIG. 4-5. The couple acting on suspended loops.

AB and CD. For the couple to produce the alignment required by the above argument, we see that the force between currents flowing in the same direction is one of attraction, and that the force between currents flowing in opposite directions is one of repulsion. (We may contrast this with the forces between electrostatic charges where likes repel and unlikes attract.) We will now deduce the magnitude of this force.

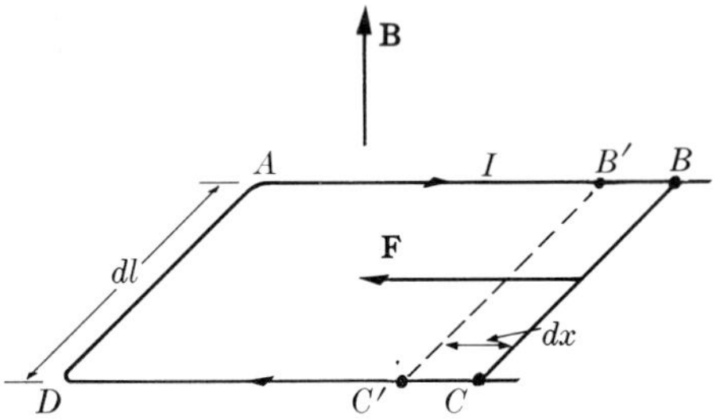

FIG. 4–6. The force acting on a current-carrying conductor in a magnetic field.

We can regard the forces on the currents as arising from the mutual coupling term, as we have seen above. The energy associated with this term is MI_1I_2. Now let us consider the case shown in Fig. 4–6. We suppose that the flux density **B** is produced by another circuit (not shown), and let the circuit in the figure be small so that **B** can be considered constant over its area A. The directions of the current and the mutual flux density **B** are related, so that the energy associated with the term MI_1I_2 is positive from the argument given in the last paragraph. This energy is given by

$$MI_1I_2 = \psi_{12}I_2 = BIA.$$

Since the potential energy of any system always tends to decrease, it follows that if the side BC were free to slide it would tend to move so as to decrease A and thereby decrease the potential energy associated with the system. Suppose it moves a distance dx to a position $B'C'$. If the force on BC is F, then the work is $F\,dx$, and this must be equal to the change in energy. Hence

$$F\,dx = BI\,dA = BI\,dl\,dx, \qquad F = BI\,dl.$$

If we tilt the circuit by rotating it about the line CD so that the flux vector **B** makes an angle θ with BC, then the change in flux would be given by $B \sin\theta\, dl\, dx$, and for the force we have

$$F = B \sin\theta\, I\, dl.$$

We see that this can be written in vector notation as

$$\mathbf{F} = I\, d\mathbf{l} \times \mathbf{B}.$$

This is a law which was originally deduced from the results of experiment, and it provides another check on our theory.

Problems

4–1 Use Ampère's law to show that the circumferential component of the field due to a current I flowing in an infinitely long straight wire is given by $I/2\pi\rho$, where ρ is the radial distance from the wire.

4–2 Calculate the field distant 1 m from the wire, given that the current is 1 amp.

4–3 Calculate the field due to a current I flowing in the wall of a cylindrical tube, assuming the wall to have negligible thickness. Consider the field both inside and outside the tube.

4–4 Verify the boundary condition for **H** in Problem 4–3.

4–5 Calculate the field due to a closely wound coil of N turns carrying current I wound on a toroid.

4–6 Given that the toroid in Problem 4–5 has a radius of 0.2 m, the turns have radius 0.01 m, and the number of turns is 100, calculate the field for a current of 1 amp in the toroid.

4–7 Verify the boundary condition for **H** in Problem 4–5.

4–8 An infinitely long solenoid closely wound on a cylinder of radius a consists of n turns per unit length carrying current I. Calculate the component of the field parallel to the axis. Draw a sketch showing how the magnitude of this field varies with distance from the axis of the cylinder.

4–9 Show that the circumferential component of the field outside the cylinder of Problem 4–8 is zero.

4–10 Use the boundary condition for **H** and the result of Problem 4–9 to calculate the circumferential component of the field just inside the coil. Obtain this component of the field at all points inside the cylinder.

4–11 Calculate the couple acting on a square coil of side a carrying a current I in the presence of a magnetic field in the plane of the coil and parallel to one side.

4–12 Calculate the magnitude of the couple in Problem 4–11, given that the coil has side 1 m and carries current 1 amp, and the field is 1 amp/m.

4–13 Calculate the couple acting on a circular coil of radius a carrying current I in the presence of a magnetic field in the plane of the coil.

4–14 The form of the results of Problems 4–11 and 4–13 suggests that the couple on a coil can be expressed in such a way that the shape of the coil need not be known. Put your results into this form.

CHAPTER 5

Maxwell's Equations

5-1 Introduction

Maxwell's equations are of fundamental importance. They describe the inherent properties of the field vectors considered in the preceding chapters and the relations between them. Furthermore, they form the starting point for the solution of many problems and have a wide variety of applications.

5-2 Faraday's law

In Section 4-2 we defined a quantity ψ by the relation

$$\psi = \int V\,dt.$$

Alternatively we may write

$$V = \frac{d\psi}{dt}.$$

We subsequently attached the significance of magnetic flux to ψ, and hence our definition is equivalent to saying that the induced voltage is equal to the rate of change of magnetic flux. This law, discovered by Faraday, is another point at which an experimental check on our theory is provided.

Using the relations of the last chapter, we can write

$$V = -\oint \mathbf{E}\cdot d\mathbf{l} = \frac{\partial\psi}{\partial t} = \frac{\partial}{\partial t}\iint \mathbf{B}\cdot d\mathbf{S}.$$

The contour integral is taken around a closed contour and the surface integral over a surface bounded by that contour. Let us apply this result to a contour of infinitesimal dimensions, obtaining

$$-\oint \mathbf{E}\cdot d\mathbf{l} = \frac{\partial}{\partial t}(\mathbf{B}\cdot\mathbf{\Delta S}).$$

We now maximize the integral by taking $\mathbf{B}$ and $\Delta\mathbf{S}$ in the same direction, divide by ΔS, and take the limit $\Delta S \to 0$:

$$-\lim_{\Delta S\to 0}\frac{(\oint \mathbf{E}\cdot d\mathbf{l})_{\max}}{\Delta S} = \frac{\partial\mathbf{B}}{\partial t}, \qquad \text{or} \qquad -\operatorname{curl}\mathbf{E} = \frac{\partial\mathbf{B}}{\partial t}.$$

This is one of Maxwell's equations.

Taking the divergence of both sides of this equation, we have

$$-\operatorname{div}\operatorname{curl}\mathbf{E} = \operatorname{div}\frac{\partial\mathbf{B}}{\partial t},$$

and since div curl is identically zero, we have

$$\operatorname{div}\frac{\partial\mathbf{B}}{\partial t} = 0.$$

Interchanging the div-operator and $\partial/\partial t$ (since div is a differential operator, this is equivalent to changing the order of differentiation, which is valid for well-behaved functions), we have

$$\frac{\partial}{\partial t}\operatorname{div}\mathbf{B} = 0.$$

Integrating this equation shows that div **B** is a constant; i.e., it is independent of time. The constant can be shown to be zero by the following argument. If we suppose that our **B**-field was produced a finite time ago, i.e., it has not always existed, then if we go back far enough in time, we shall have $\mathbf{B} = 0$ and therefore $\operatorname{div}\mathbf{B} = 0$. Since we have proved quite generally that div **B** is independent of time, the process of changing **B** from zero to some finite value cannot change div **B**. Hence we have the relation

$$\operatorname{div}\mathbf{B} = 0.$$

Thus **B** is a field which has no sources. In the last chapter, we encountered the difficulty of terminating the **B** lines, and we overcame it there by closing the lines upon themselves. The result above shows that this is the only solution. If **B** has no sources, then the lines can never terminate and must always close upon themselves, a fact which can be verified experimentally. The **D**-lines terminate on electric charges. Experiment has failed to produce any evidence of free magnetic charge, and hence we would expect div **B** to be zero.

The results of this section enable us to complete the derivation of the boundary conditions for the field vectors. In Section 3–7 we showed that there is a discontinuity in the normal component of **D** at a boundary equal to the surface charge density at the boundary, and that the tangential component of **E** is continuous at the boundary. The second relation was proved only for static fields. Its extension to time-varying fields is achieved by using the relation

$$\oint \mathbf{E}\cdot d\mathbf{l} = -\frac{\partial}{\partial t}\iint \mathbf{B}\cdot d\mathbf{S}.$$

Referring back to the proof in Section 3–7, we take the surface integral over the area $ABCD$. Since this is an infinitesimal area, we can suppose that **B** is

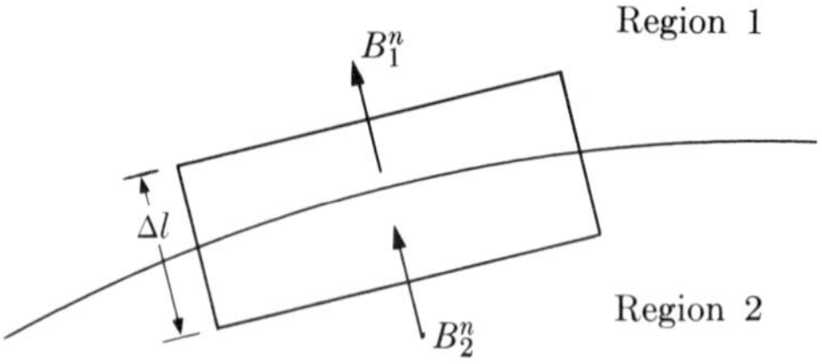

FIG. 5–1. The boundary condition for **B**.

constant over it, and the integral becomes $\mathbf{B} \cdot \mathbf{\Delta S}$, where $\mathbf{\Delta S}$ is the area $ABCD$. However, when we make BC and DA tend to zero, the area also becomes zero, and so the right-hand side of the above equation is zero. The rest of the proof follows as before.

For the magnetic field, we showed in Section 4–6 how the **H**-vector varies at a boundary. In the presence of a current sheet at the boundary, there is a discontinuity in the tangential component of **H** perpendicular to the current, whose magnitude is equal to the surface current density. To complete the picture we need the condition for **B** at a boundary. This is obtained from the relation $\text{div}\, \mathbf{B} = 0$. Consider a cylindrical surface across the boundary with cross-sectional area ΔS, length Δl, and axis perpendicular to the boundary. If we integrate $\text{div}\, \mathbf{B}$ throughout the volume and apply the divergence theorem, we have $\oiint \mathbf{B} \cdot d\mathbf{S} = 0$. The contribution from the end of the cylinder in Region 1 is $B_1^n \, \Delta S$, where B_1^n is the normal component of **B**. The contribution from the end in Region 2 is $-B_2^n \, \Delta S$, the negative sign arising because, if the two vectors are pointing in the same direction, B_1^n will give a flux out of the volume element and B_2^n a flux into it (Fig. 5–1). To get the difference between B_1^n and B_2^n at the boundary we must let $\Delta l \to 0$, so that B_1^n is at the boundary in Region 1 and B_2^n at the boundary in Region 2. The contribution to the integral from the curved surface of the cylinder then vanishes, and the integral becomes $(B_1^n - B_2^n)\, \Delta S$. Hence we have

$$(B_1^n - B_2^n)\, \Delta S = 0, \qquad \text{or} \qquad B_1^n = B_2^n.$$

The normal component of **B** is continuous at the boundary.

5–3 Maxwell's displacement current

In the last chapter we derived from Ampère's law the relation $\text{curl}\, \mathbf{H} = \mathbf{J}$. If we take the divergence of this equation, we get $\text{div}\, \mathbf{J} = 0$. This is not true in general. The equation of continuity which we derived in Chapter 1 gave the result

$$\text{div}\, \mathbf{J} = -\frac{\partial q_v}{\partial t}.$$

We also know that $\text{div}\, \mathbf{D} = q_v$, and hence $\mathbf{J}$ and $\partial \mathbf{D}/\partial t$ have the same dimensions. Maxwell added the term $\partial \mathbf{D}/\partial t$ to the above equation and called this term the *displacement current.* Thus we have

$$\text{curl}\,\mathbf{H} = \frac{\partial \mathbf{D}}{\partial t} + \mathbf{J}.$$

Taking the divergence of both sides, we obtain

$$\frac{\partial}{\partial t}\,\text{div}\,\mathbf{D} + \text{div}\,\mathbf{J} = 0, \qquad \text{or} \qquad \text{div}\,\mathbf{J} = -\frac{\partial q_v}{\partial t}.$$

We noted in the last chapter that Ampère's law might hold only for the static case. We see that this is in fact true. The term added by Maxwell is zero in the static case, since $\partial/\partial t = 0$. The relation therefore reduces to Ampère's original form of the equation. The term added by Maxwell makes the equation consistent with the equation of continuity for time-varying fields.

5-4 Maxwell's equations

We have now obtained the four equations which comprise Maxwell's equations. They are

$$-\text{curl}\,\mathbf{E} = \frac{\partial \mathbf{B}}{\partial t}, \qquad \text{curl}\,\mathbf{H} = \frac{\partial \mathbf{D}}{\partial t} + \mathbf{J},$$

$$\text{div}\,\mathbf{D} = q_v, \qquad \text{div}\,\mathbf{B} = 0.$$

These four equations contain five vector fields, **E**, **D**, **B**, **H**, and **J**, and one scalar field, q_v. Another equation is provided by the equation of continuity $\text{div}\,\mathbf{J} = -\partial q_v/\partial t$. However, we may note again that these equations are not independent. Eliminating q_v between the two equations containing it gives

$$\frac{\partial}{\partial t}\,\text{div}\,\mathbf{D} + \text{div}\,\mathbf{J} = 0,$$

a result implied by the curl **H** relation and readily obtained from it by taking the divergence of both sides. We also saw that $\text{div}\,\mathbf{B} = 0$ is implied by the curl **E** equation. Thus the two curl equations provide two independent equations among five unknowns.

5-5 The constitutive relations

To be able to solve the above equations for the unknowns, we need three more equations. In the previous chapters, we deduced additional relations between the field vectors, and the three required are

$$\mathbf{J} = \sigma\mathbf{E}, \qquad \mathbf{D} = \epsilon\mathbf{E}, \qquad \mathbf{B} = \mu\mathbf{H}.$$

These contain the parameters σ, ϵ, and μ which depend on the medium in which the fields exist. The expressions describe the constitution of the medium (so far as the fields are concerned) and thus are referred to as the *constitutive relations.*

5-6 Boundary conditions

The two curl equations are differential equations in four variables (x, y, z, and t). We are accustomed to the fact that when we integrate an ordinary differential equation an arbitrary constant occurs which must be determined by some additional given condition. For example, in this case we have a partial differential with respect to time, and the constant introduced here might be determined by the value at $t = 0$ or the initial conditions of the system. The integration over the space variables will require similar additional information. This is usually obtained by specifying the value of the field over a surface bounding the region of interest and is thus often referred to as the *boundary condition.*

Problems

5-1 By integrating Maxwell's equations over an area and using Stokes' theorem, obtain the integral form of the equations. [*Hint:* This process is similar to the integration of div $\mathbf{D} = q_v$ to obtain Gauss' theorem.]

5-2 If the permittivity is proportional to the electric field, what is the corresponding constitutive relation? How is **D** affected if **E** is doubled?

5-3 If the permeability is proportional to the first derivative of the magnetic field with respect to time, what is the corresponding constitutive relation? How is **B** affected if **H** is doubled?

5-4 If the electric flux density is proportional to the second derivative of the field with respect to time, what is the corresponding constitutive relation? How is **D** affected if **E** is doubled?

5-5 Use the divergence and curl properties of the magnetic field to show that the radial component of the field due to an infinite solenoid (Problem 4-6) is zero.

5-6 A closely wound coil of N turns carrying current I is wound on a toroidal core which has a small gap of width Δ. The permeability of the core material is μ, and the whole assembly is situated in a vacuum. Calculate **B** and **H** at all points.

5-7 In Problem 5-6 calculate the field in the gap for a coil of 100 turns, the radius of the toroid being 0.1 m, the radius of the turns 0.01 m, the width of the gap 0.5 cm, the relative permeability of the core 1000, and the current 1 amp.

CHAPTER 6

Static Fields

6–1 Introduction

In this chapter we shall use the results developed previously to solve problems for the case in which the fields are static; i.e., they are not functions of time. Thus in our equations we can set $\partial/\partial t = 0$. We shall investigate electrostatic and magnetostatic fields.

6–2 Maxwell's equations

If we take Maxwell's equations and put $\partial/\partial t = 0$, then we obtain

$$\text{curl}\,\mathbf{E} = 0, \qquad \text{curl}\,\mathbf{H} = \mathbf{J},$$

$$\text{div}\,\mathbf{D} = q_v, \qquad \text{div}\,\mathbf{B} = 0.$$

We see that for this case the equations separate into two independent sets. The first contains the three variables **E**, **D**, and q_v, which are the electric quantities, and the second contains **H**, **B**, and **J**, which are magnetic quantities. For the special case of static fields, the electric and magnetic fields are independent. This is not the case if the fields are time dependent, as Maxwell's equations show.

6–3 Electrostatic fields

The equations of electrostatics are, of course, already familiar. We note that we have only two equations among three unknowns. The third, as we saw in the last chapter, comes from the constitutive relations, and the appropriate equation for this case is $\mathbf{D} = \epsilon\mathbf{E}$.

The first equation shows that **E** is a conservative field and can be written as the gradient of a scalar field. (The operation curl grad is identically zero.) So we can write $\mathbf{E} = -\text{grad}\,V$. The minus sign is conventional and is in agreement with our previous work. From the other two equations we obtain

$$\text{div}\,\epsilon\mathbf{E} = q_v.$$

The divergence operator is a space derivative; it contains the operators $\partial/\partial x$, $\partial/\partial y$, and $\partial/\partial z$. If we suppose that ϵ is not a function of x, y, or z, then we can bring it outside the operator and write

$$\text{div}\,\epsilon\mathbf{E} = \epsilon\,\text{div}\,\mathbf{E}.$$

The physical significance of this statement is that the permittivity is not a function of position; it has the same value at all points throughout the medium. Such a medium is said to be homogeneous. Continuing the development of the equations, we have

$$\epsilon \operatorname{div} \mathbf{E} = q_v, \qquad \operatorname{div} \mathbf{E} = \frac{q_v}{\epsilon}, \qquad \operatorname{div} \operatorname{grad} V = -\frac{q_v}{\epsilon},$$

or, using the ∇-operator, we obtain

$$\nabla^2 V = -\frac{q_v}{\epsilon},$$

which is Poisson's equation. For charge-free regions, i.e., where $q_v = 0$, we have Laplace's equation,

$$\nabla^2 V = 0.$$

We can now solve for the field or potential for any given charge distribution by solving Poisson's equation, bearing in mind that we have to supply some additional information (boundary conditions) when the equation is integrated. One method of solving this equation actually amounts to reversing the procedure we have used above to obtain the equation. We shall now carry out this procedure, developing various techniques for solving static field problems on the way.

6-4 The solution of Poisson's equation

Starting from Poisson's equation and substituting $\mathbf{E} = -\operatorname{grad} V$, we have

$$\operatorname{div} \mathbf{E} = \frac{1}{\epsilon} q_v.$$

Integrating this throughout a volume and transforming the left-hand side by the divergence theorem gives

$$\oiint \mathbf{E} \cdot d\mathbf{S} = \frac{1}{\epsilon} \iiint q_v \, d\tau,$$

where the surface integral is taken over the surface bounding the volume through which the volume integral is taken. This result, which we recognize as Gauss' theorem (Section 3-6), provides a valuable method of solving for the field when the integrals can be readily obtained, as is the case for fields possessing a high degree of symmetry. As an example, let us take the field of a point charge. From the symmetry of this case we see that the field will be directed radially, and its magnitude will depend only upon the distance from the charge. Thus the field over the surface of a sphere with radius r centered at the charge will have constant magnitude and will always be perpendicular to the surface. Hence the surface integral is readily obtained as

$$\oiint \mathbf{E} \cdot d\mathbf{S} = 4\pi r^2 E.$$

The volume integral is also easily obtained, since the charge density is zero everywhere inside the sphere except at the center where there is a point charge of magnitude Q. Hence

$$4\pi r^2 E = \frac{1}{\epsilon} Q, \qquad \text{or} \qquad E = \frac{Q}{4\pi\epsilon r^2}.$$

We have already integrated a differential equation once, apparently without supplying the constant of integration. In fact, any constant **E**-vector could be added to our solution, and the resulting field would still satisfy Poisson's equation. However, if we specify that the field must be zero as r approaches infinity, then this constant must be zero. Note that specifying a zero field at infinity specifies the field over the boundary surface and constitutes a boundary condition.

Having obtained **E**, we can calculate V from the relation

$$V = -\int \mathbf{E} \cdot d\mathbf{l}.$$

Let us consider the integration taken along a radial line drawn from the point charge so that **E** and $d\mathbf{l}$ are in the same direction. Then

$$V = -\int \frac{Q}{4\pi\epsilon r^2}\, dr = \frac{Q}{4\pi\epsilon r} + C,$$

where C is the constant of integration. Again we must specify a boundary condition, and we suppose that the potential V is also zero at infinity. Hence

$$V = \frac{Q}{4\pi\epsilon r}.$$

We have now obtained the field and the potential due to a point charge, and may use the principle of superposition to obtain the field due to a distribution of point charges (and in the limit of a continuous charge distribution) provided the medium is a linear one. We note that the expression for **E** contains an ϵ in the denominator, and thus if the field at a point due to one charge is to be independent of the field due to all the other charges, then ϵ must be independent of **E**. This implies that the constitutive relation is linear. (Suppose, for example, that ϵ were a function of E and could be written $\epsilon_1 + \epsilon_2 E$, where ϵ_1 and ϵ_2 are constants; then the constitutive relation would become $D = \epsilon_1 E + \epsilon_2 E^2$, which is obviously nonlinear.) Hence a linear electric medium is one whose permittivity is not a function of the impressed electric field. (Compare this with the linear resistor which has a resistance that is not a function of the voltage across it.)

If we were to calculate the field at a point due to each of the given charges, then the total field would be their sum. Note, however, that this would be a vector sum. A simpler procedure is to calculate the potential, which involves a scalar summation (potential being a scalar), and then calculate $-\text{grad}\, V$.

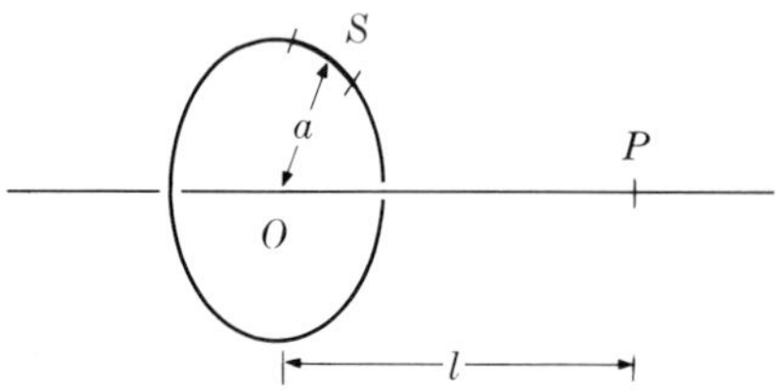

FIG. 6-1. The ring charge.

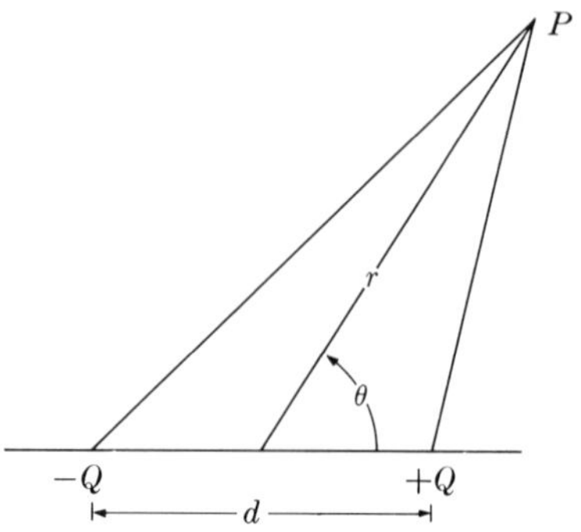

FIG. 6-2. The dipole.

6-5 Examples of static fields

We illustrate the procedure outlined above by considering a ring of charge and calculating the field on the axis (Fig. 6-1). Let us consider an element of charge situated at S. If the charge per unit length of the ring is q, and the element subtends an angle $d\alpha$ at the center of the ring, then the charge at S is $qa\,d\alpha$. The potential at P due to this element is

$$dV = \frac{qa\,d\alpha}{4\pi\epsilon SP} = \frac{qa\,d\alpha}{4\pi\epsilon\sqrt{a^2 + l^2}},$$

and the potential due to the whole ring is arrived at by integrating from zero to 2π. None of the quantities in the expression is a function of α, and thus

$$V = \frac{2\pi aq}{4\pi\epsilon\sqrt{a^2 + l^2}} = \frac{Q}{4\pi\epsilon\sqrt{a^2 + l^2}}.$$

From the symmetry of this problem we see that the direction of the field at P will be along the axis, and its magnitude will be given by $-(dV/dl)$; hence

$$E = -\frac{dV}{dl} = \frac{Q}{4\pi\epsilon}\frac{2l}{(a^2 + l^2)^{3/2}}.$$

Note that there are two different types of operations involved in this calculation. We first calculate the potential at the point P due to a charge element on the ring, and next integrate over all the charges. Since this is a simple problem, there is no need to specify the position of S in terms of coordinates. In more complicated problems this would be necessary, and the integration would be made with respect to those coordinates specifying the position of the sources. The integration, which in this case was elementary, gave us the potential at P in terms of the distance l from the origin, that is, we obtain potential as a function of position. The second operation, that of calculating the gradient, involves an operation on this function, and we differentiate with respect to the coordinate specifying the position of the point at which we have the potential, namely, l.

As a second example, we consider a dipole. Two equal and opposite charges of magnitude Q are separated by a distance d. We let $d \to 0$ and $Q \to \infty$ in such a way that $Qd = P$, say, remains constant. The result is a dipole of moment P. We see that this problem has rotational symmetry about the line joining the two charges and hence may be solved as a two-dimensional problem. If we calculate the equipotential lines in the plane of the page (Fig. 6-2), we obtain the equipotential surfaces by rotating this pattern about the line joining Q and $-Q$. The potential at P is given by

$$V = \frac{Q}{4\pi\epsilon}\left(\frac{1}{r - \frac{1}{2}d\cos\theta} - \frac{1}{r + \frac{1}{2}d\cos\theta}\right).$$

Since d is infinitesimal for the dipole, then $d \ll r$. Expanding the terms in parentheses and neglecting the higher-order terms, we have

$$V = \frac{Q}{4\pi\epsilon r}\left[\left(1 + \frac{d}{2r}\cos\theta\right) - \left(1 - \frac{d}{2r}\cos\theta\right)\right] = \frac{Q\,d\cos\theta}{4\pi\epsilon r^2}$$

$$= \frac{P\cos\theta}{4\pi\epsilon r^2}.$$

If we define P as a vector whose direction lies along the line joining the two charges and pointing toward the positive charge, we can write

$$V = \frac{\mathbf{P}\cdot\mathbf{u}_r}{4\pi\epsilon r^2}.$$

To obtain the field we must calculate the gradient of this quantity. We know that the gradient has no component perpendicular to the page, since this direction lies in the equipotential surface. The gradient vector therefore lies in the plane of the page. Since the potential is expressed in terms of the plane polar coordinates r and θ, it will be convenient to calculate the gradient in these terms. From Chapter 1 we have

$$\text{grad}_l\,V = \lim_{\Delta l \to 0}\frac{\Delta V}{\Delta l},$$

where Δl is the element of length measured in the l-direction. From Fig. 6-3 we see that the element of length measured in the direction of increasing r is Δr,

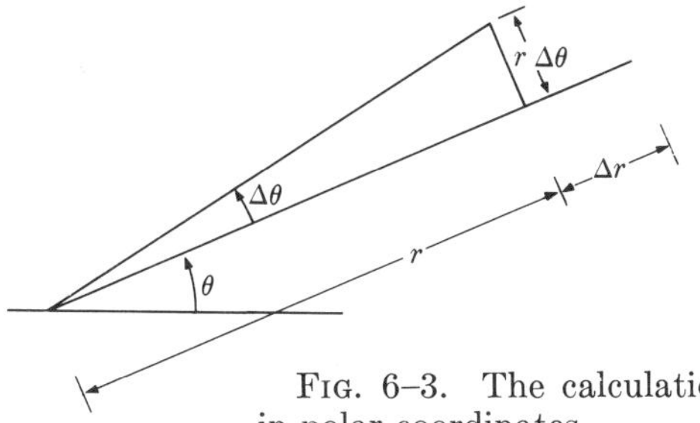

FIG. 6-3. The calculation of gradient in polar coordinates.

and in the direction of increasing θ is $r\,\Delta\theta$. (Note that both of these have the dimensions of length.) Hence

$$\text{grad}_r\, V = \frac{\partial V}{\partial r}, \qquad \text{grad}_\theta\, V = \frac{1}{r}\frac{\partial V}{\partial \theta}.$$

Thus for the field components we derive

$$E_r = -\frac{P\cos\theta}{4\pi\epsilon}\left(-\frac{2}{r^3}\right) = \frac{2P\cos\theta}{4\pi\epsilon r^3},$$

$$E_\theta = -\frac{P}{4\pi\epsilon r^2}\frac{(-\sin\theta)}{r} = \frac{P\sin\theta}{4\pi\epsilon r^3}.$$

We may therefore write the field as follows:

$$\mathbf{E} = \frac{P}{4\pi\epsilon r^3}(\mathbf{u}_r\, 2\cos\theta + \mathbf{u}_\theta \sin\theta).$$

In the first example above we obtained the field using Gauss' theorem, and in the second example we used the potential integral method. Finally we take an example to show how Poisson's equation can be integrated directly.

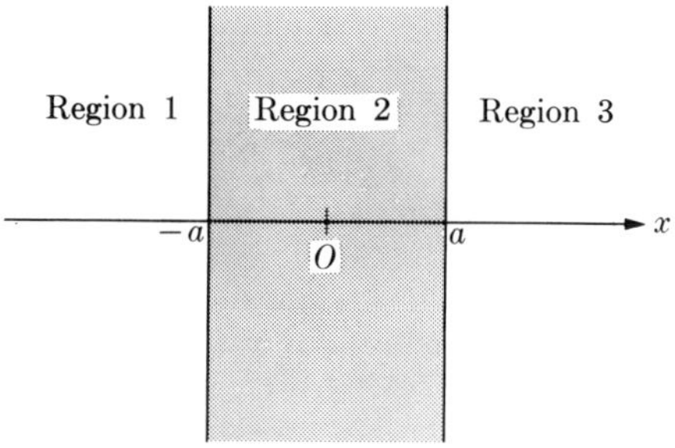

FIG. 6–4. A slab charge.

Let us consider a charge distribution given by q_v for $-a < x < a$ and zero everywhere else. This constitutes an infinite slab of charge of thickness $2a$ parallel to the yz-plane. The symmetry of the problem indicates that the potential and the field will not depend upon the y- or z-coordinates, and so $\partial/\partial y = \partial/\partial z = 0$, and Poisson's equation becomes

$$\frac{d^2V}{dx^2} = \begin{cases} -\dfrac{q_v}{\epsilon}, & -a < x < a, \\ 0 & \text{elsewhere.} \end{cases}$$

We see from Fig. 6–4 that this is a three-region problem, the first region being $x < -a$, the second $-a < x < a$, and the third $x > a$. We would expect the potential to depend only on the distance from the slab; that is, V is an even function of x. The magnitude of the field will also depend only on the distance from the slab, but it will be in opposite directions on opposite sides of the slab. Hence E is to be an odd function of x. Integrating Poisson's equation once for the three regions gives

$$\frac{dV}{dx} = K_1, \qquad \frac{dV}{dx} = -\frac{q_v}{\epsilon}x + K_2, \qquad \frac{dV}{dx} = K_3.$$

In addition, $dV/dx = -E$, and this must be an odd function. Hence $K_1 = -K_3$, and $K_2 = 0$. Integrating again, we have

$$V = K_1x + C_1, \qquad V = -\frac{q_v}{2\epsilon}x^2 + C_2, \qquad V = -K_1x + C_3.$$

This must be an even function, and so $C_1 = C_3$. We can specify the zero of potential arbitrarily, and it is convenient to take $V = 0$ for $x = 0$. Hence $C_2 = 0$, and the equations reduce to

$$V = K_1x + C_1, \qquad V = -\frac{q_v}{2\epsilon}x^2, \qquad V = -K_1x + C_1.$$

Next let us determine K_1 and C_1. The field and the potential must be continuous at $x = a$ and $x = -a$ since there are no surface charges. At $x = -a$ the field is

$$K_1 = \frac{q_v}{\epsilon}a.$$

(We obtain the same result at $x = a$ since we constructed the solutions to have the required symmetry.) For the potential we have

$$-K_1a + C_1 = -\frac{q_v}{2\epsilon}a^2,$$

$$\therefore C_1 = \frac{q_va^2}{2\epsilon}.$$

Thus for the three regions,

$$V = \frac{q_v}{\epsilon}ax + \frac{q_v}{2\epsilon}a^2, \qquad E = -\frac{q_v}{\epsilon}a,$$

$$V = -\frac{q_v}{2\epsilon}x^2, \qquad E = \frac{q_v}{\epsilon}x,$$

$$V = -\frac{q_v}{\epsilon}ax + \frac{q_v}{2\epsilon}a^2, \qquad E = \frac{q_v}{\epsilon}a.$$

These results are plotted in Fig. 6–5.

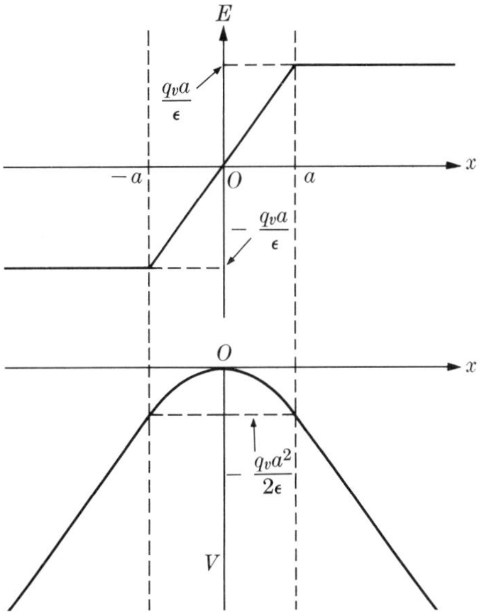

FIG. 6–5. The field and potential distribution of the slab charge.

6-6 Magnetostatic fields

We will carry through an analogous procedure for the magnetic case. The fundamental equations are

$$\text{curl } \mathbf{H} = \mathbf{J}, \qquad \text{div } \mathbf{B} = 0, \qquad \mathbf{B} = \mu\mathbf{H}.$$

For a homogeneous medium we can write

$$\text{curl } \mathbf{B} = \text{curl } \mu\mathbf{H} = \mu \text{ curl } \mathbf{H},$$

and hence

$$\text{curl } \mathbf{B} = \mu\mathbf{J}.$$

Since the operation div curl is identically equal to zero, we can write **B** as the curl of another vector field **A**, namely,

$$\mathbf{B} = \text{curl } \mathbf{A}.$$

The second relation is then automatically satisfied, and from the first we have

$$\text{curl curl } \mathbf{A} = \mu\mathbf{J}.$$

The operation curl curl will arise frequently in field theory, and so we shall investigate it further. Writing curl **A** in Cartesian coordinates, we have

$$\text{curl } \mathbf{A} = \mathbf{u}_x\left(\frac{\partial A_z}{\partial y} - \frac{\partial A_y}{\partial z}\right) + \mathbf{u}_y\left(\frac{\partial A_x}{\partial z} - \frac{\partial A_z}{\partial x}\right) + \mathbf{u}_z\left(\frac{\partial A_y}{\partial x} - \frac{\partial A_x}{\partial y}\right).$$

To avoid a long, cumbersome expression, consider the x-component of the curl of curl **A**, namely, curl_x (curl **A**). This may be written

$$\begin{aligned}\text{curl}_x (\text{curl } \mathbf{A}) &= \frac{\partial}{\partial y}(\text{curl}_z \mathbf{A}) - \frac{\partial}{\partial z}(\text{curl}_y \mathbf{A}) \\ &= \frac{\partial}{\partial y}\left(\frac{\partial A_y}{\partial x} - \frac{\partial A_x}{\partial y}\right) - \frac{\partial}{\partial z}\left(\frac{\partial A_x}{\partial z} - \frac{\partial A_z}{\partial x}\right) \\ &= \frac{\partial}{\partial x}\left(\frac{\partial A_x}{\partial x} + \frac{\partial A_y}{\partial y} + \frac{\partial A_z}{\partial z}\right) - \left(\frac{\partial^2 A_x}{\partial x^2} + \frac{\partial^2 A_x}{\partial y^2} + \frac{\partial^2 A_x}{\partial z^2}\right),\end{aligned}$$

where we have added the term $\partial^2 A_x/\partial x^2$ to the first parentheses and subtracted it again in the second. Thus we can write

$$\text{curl}_x (\text{curl } \mathbf{A}) = \frac{\partial}{\partial x}(\text{div } \mathbf{A}) - \nabla^2 A_x,$$

or

$$\text{curl}_x (\text{curl } \mathbf{A}) = \text{grad}_x (\text{div } \mathbf{A}) - \nabla^2 A_x.$$

Similarly, it may be shown that

$$\text{curl}_y\,(\text{curl}\,\mathbf{A}) = \text{grad}_y\,(\text{div}\,\mathbf{A}) - \nabla^2 A_y,$$

and

$$\text{curl}_z\,(\text{curl}\,\mathbf{A}) = \text{grad}_z\,(\text{div}\,\mathbf{A}) - \nabla^2 A_z.$$

Multiplying by the unit vectors and adding, we obtain

$$\text{curl curl}\,\mathbf{A} = \text{grad div}\,\mathbf{A} - (\mathbf{u}_x\nabla^2 A_x + \mathbf{u}_y\nabla^2 A_y + \mathbf{u}_z\nabla^2 A_z).$$

Note that A_x, A_y, and A_z are scalars (they are the *magnitudes* of the x-, y-, and z-components of the vector), and that the operation ∇^2 is made on a scalar; it is the operation div grad, and the grad operator operates on a scalar field. As a shorter way of writing this equation we define the operation ∇^2 on a vector by the equation

$$\nabla^2\mathbf{A} = \mathbf{u}_x\nabla^2 A_x + \mathbf{u}_y\nabla^2 A_y + \mathbf{u}_z\nabla^2 A_z.$$

This merely means that we write the term in parentheses symbolically as $\nabla^2\mathbf{A}$. Note also that this operator has been defined here only in Cartesian components. Thus we have the relation curl curl $\mathbf{A}$ = grad div $\mathbf{A} - \nabla^2\mathbf{A}$.

This is, of course, true for any general vector **A**. We now return to the vector **A** which we introduced above by the relation $\mathbf{B} = \text{curl}\,\mathbf{A}$. We have seen that, in general, a vector field can have both sources and vortexes, which are independent of each other. So far we have specified only the vortexes of our **A**-field. To complete the definition of **A** we must also specify its sources, which we now choose to be zero everywhere. Thus div $\mathbf{A} = 0$. With this choice of vector field **A**, we have

$$\text{curl curl}\,\mathbf{A} = -\nabla^2\mathbf{A} = \mu\mathbf{J}, \qquad \text{or} \qquad \nabla^2\mathbf{A} = -\mu\mathbf{J}.$$

From our definition of $\nabla^2\mathbf{A}$ we interpret this equation to mean

$$\nabla^2 A_x = -\mu J_x, \qquad \nabla^2 A_y = -\mu J_y, \qquad \nabla^2 A_z = -\mu J_z.$$

Because of its similarity to Poisson's equation, $\nabla^2\mathbf{A} = -\mu\mathbf{J}$ is known as the *vector Poisson equation*, and the vector **A**, again by analogy, is known as the *vector potential.*

6–7 The solution of the vector Poisson equation

Reverting to the electrostatic case for the moment, we have

$$\nabla^2 V = -\frac{1}{\epsilon}\,q_v.$$

We saw that the potential at a distance l from a charge Q could be written $Q/4\pi\epsilon l$. We can write the point charge in terms of a charge density q_v and an infinitesimal volume element $d\tau$, the point charge then being $q_v d\tau$ and the potential $q_v\, d\tau/4\pi\epsilon l$. Hence for a charge distribution we obtain the potential at a point by adding all the contributions from the charge elements $q_v\, d\tau$. In other words, we calculate

$$V = \iiint \frac{q_v\, d\tau}{4\pi\epsilon l},$$

where the integral is taken throughout the volume containing the charge density q_v. Hence the solution to the equation

$$\nabla^2 V = -\frac{1}{\epsilon} q_v$$

is

$$V = \iiint \frac{q_v\, d\tau}{4\pi\epsilon l}.$$

By merely changing the symbols, the solution to

$$\nabla^2 A_x = -\mu J_x$$

must be

$$A_x = \iiint \frac{\mu J_x\, d\tau}{4\pi l},$$

and similarly for A_y and A_z. We may write this equation in vector form as

$$\mathbf{A} = \iiint \frac{\mu \mathbf{J}\, d\tau}{4\pi l}.$$

This is the potential integral solution for the magnetic case. Again, taking a hint from the electrostatic case, we might proceed as follows:

$$\nabla^2 \mathbf{A} = -\mu \mathbf{J},$$

and since div $\mathbf{A} = 0$, we have

$$\text{grad div } \mathbf{A} - \nabla^2 A = \mu \mathbf{J},$$

$$\therefore \text{curl curl } \mathbf{A} = \mu \mathbf{J}.$$

But

$$\text{curl } \mathbf{A} = \mathbf{B} = \mu \mathbf{H},$$

$$\therefore \text{curl } \mu\, \mathbf{H} = \mu \text{ curl } \mathbf{H} = \mu \mathbf{J},$$

$$\therefore \text{curl } \mathbf{H} = \mathbf{J}.$$

Integrating this over an area and transforming by Stokes' theorem, we have

$$\oint \mathbf{H} \cdot d\mathbf{l} = \oiint \mathbf{J} \cdot d\mathbf{S},$$

which is Ampère's law. This is the relation which corresponds to Gauss' theorem in electrostatics; it is useful if the problem has a high degree of symmetry. We now have the three methods of solving the magnetic case: the potential integral, Ampère's law, and direct integration, which correspond closely to the three methods developed for electrostatics.

6–8 Examples of magnetostatic fields

Let us consider the magnetic field due to an infinite line current. This problem has cylindrical symmetry; the value of the field depends only on the distance from the line, and since we know that H-lines must be closed curves, it follows that they must be circles centered on the line (Fig. 6–6). We now apply Ampère's theorem to a contour consisting of a circle centered on the line. The vectors $\mathbf{H}$ and $d\mathbf{l}$ are therefore in the same direction, and H has constant magnitude around the contour. Hence $\oint \mathbf{H} \cdot d\mathbf{l} = 2\pi\rho H$, where ρ is the radius of the circle and H is the magnitude of the field at a distance ρ from the wire. The integral $\oiint \mathbf{J} \cdot d\mathbf{S}$ is just I, the current passing through the contour. Hence

$$H = \frac{I}{2\pi\rho}.$$

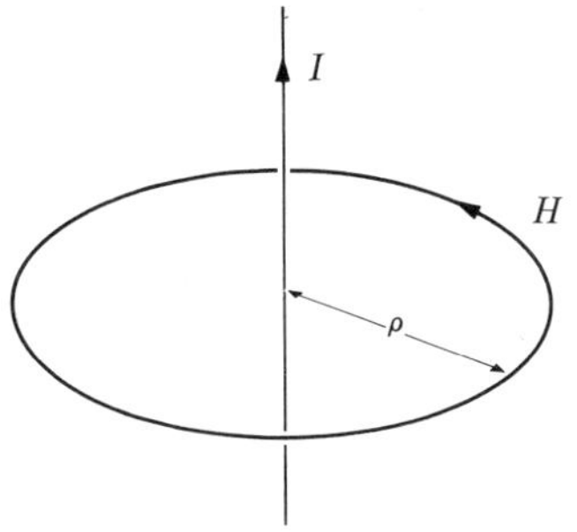

FIG. 6–6. An infinite line current and its magnetic field.

Now let us consider an element of current situated at the origin, i.e., a current of magnitude I flowing in a wire of length dl situated at the origin. (This is not, of course, a physically realizable case since current always flows in closed loops, but it can be solved mathematically if we postulate such a current.) We solve this problem by means of the potential integral. If $d\mathbf{l}$ is directed along the z-axis, then there are no currents anywhere in the system flowing in the x- or y-directions. Hence $J_x = J_y = 0$, and from the potential integrals, $A_x = A_y = 0$. For A_z we have

$$A_z = \iiint \frac{\mu J_z \, d\tau}{4\pi\, d}.$$

Since $J_z = 0$ everywhere except at the origin, then (Fig. 6–7a)

$$A_z = \frac{\mu}{4\pi OP} \iiint J_z \, d\tau, \qquad \iiint J_z \, d\tau = \iiint J_z \, dx \, dy \, dz, \qquad \iint J_z \, dx \, dy = I,$$

where I is the current flowing in the z-direction. But this is zero for all values of z except over the element dl situated at the origin, and hence

$$\iiint J_z \, d\tau = \int I \, dz = I \, dl.$$

If we let $OP = r$, we have

$$A_z = \frac{\mu I\,dl}{4\pi r}.$$

To obtain the **B**-field we calculate curl **A**. Since **A** has only a z-component, we have

$$\text{curl}\,\mathbf{A} = \mathbf{u}_x \frac{\partial A_z}{\partial y} - \mathbf{u}_y \frac{\partial A_z}{\partial x}.$$

In Cartesian coordinates, A_z is given by

$$A_z = \frac{\mu I\,dl}{4\pi} \frac{1}{\sqrt{x^2 + y^2 + z^2}},$$

$$\therefore\ \text{curl}\,\mathbf{A} = \frac{\mu I\,dl}{4\pi r^3}(-\mathbf{u}_x y + \mathbf{u}_y x).$$

Thus the **B**-vector has no z-component. It is constructed in Fig. 6–7(b), and from the geometry one can see that its direction is perpendicular to ON. This is also shown in Fig. 6–7(a). The magnitude of the vector in parentheses is $\sqrt{x^2 + y^2} = ON$, and from Fig. 6–7(a), $ON/r = \sin\theta$. Thus

$$B = |\text{curl}\,\mathbf{A}| = \frac{\mu I\,dl}{4\pi} \frac{1}{r^2} \frac{ON}{r} = \frac{\mu I\,dl \sin\theta}{4\pi r^2}.$$

We note that the direction of the vector **B** is perpendicular to ON and to O_z. This means that it is perpendicular to the plane $OzPN$, and its direction is therefore defined uniquely by specifying that it is perpendicular to Oz and to OP (i.e., to the vectors $d\mathbf{l}$ and $\mathbf{r}$, where $\mathbf{r}$ is the vector length drawn from the current element to the field point concerned). Furthermore, the magnitude is proportional to dl and to $\sin\theta$, which is the angle between the vectors $d\mathbf{l}$ and $\mathbf{r}$.

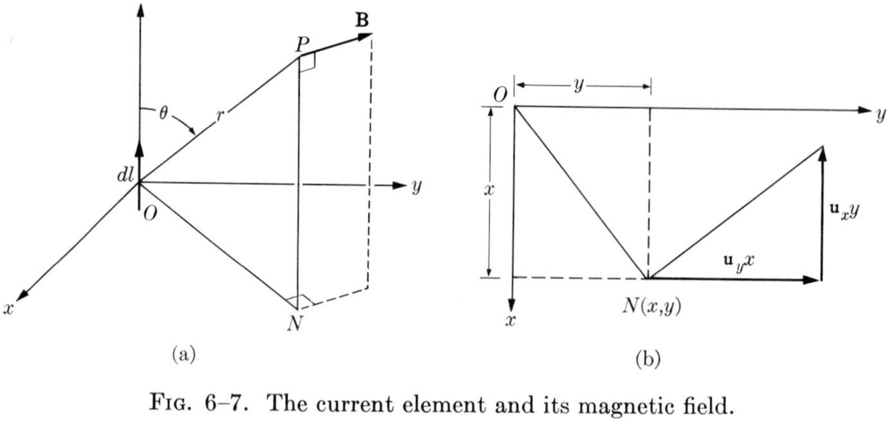

Fig. 6–7. The current element and its magnetic field.

Hence it may be written as a vector product,

$$\mathbf{B} = \frac{\mu I \, d\mathbf{l} \times \mathbf{u}_r}{4\pi r^2}, \qquad \text{and} \qquad \mathbf{H} = \frac{I \, d\mathbf{l} \times \mathbf{u}_r}{4\pi r^2}.$$

The latter expression is also commonly written as

$$\mathbf{H} = \frac{I \, d\mathbf{l} \times \mathbf{r}}{4\pi r^3}.$$

This is a relation known as the Biot-Savart law, which was discovered by the two scientists in the course of carrying out experiments on magnetic fields produced by electric currents. Here we have another important experimental check of theoretical results. The law derived from our theory agrees with that deduced from experimental observations, and hence the theory has been verified.

We saw in Section 4–7 that the force on a current element is given by

$$\mathbf{F} = I \, d\mathbf{l} \times \mathbf{B}.$$

If $\mathbf{B}$ is due to a second current element $d\mathbf{l}_2$ carrying current I_2, then

$$\mathbf{B} = \frac{\mu I_2 \, d\mathbf{l}_2 \times \mathbf{r}}{4\pi r^3}.$$

If we denote by $d\mathbf{l}_1$ the original current element, carrying current I_1, then the force between the two current elements is

$$\mathbf{F} = \frac{\mu I_1 I_2 \, d\mathbf{l}_1 \times (d\mathbf{l}_2 \times \mathbf{r})}{4\pi r^3}.$$

This is Ampère's law, which was deduced from the results of a series of experiments on the forces between current-carrying conductors and which provides another experimental check of the results of our theory. This law corresponds to Coulomb's law in electrostatics and forms the starting point for some classical treatments of the subject.

The Biot-Savart law can be applied to the calculation of fields. Let us consider the field at the center of a ring of radius a carrying a current I (Fig. 6–8). By this law the field due to an element of the ring $a \, d\alpha$ is given by

$$dH = \frac{Ia \, d\alpha}{4\pi a^2}.$$

For the field due to the complete ring we integrate α from zero to 2π, obtaining

$$H = \frac{I}{2a}.$$

The direction of H is along the axis, as shown in the figure.

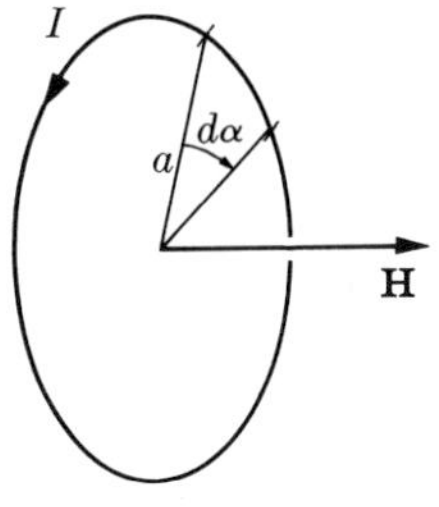

FIG. 6–8. The ring current and its magnetic field.

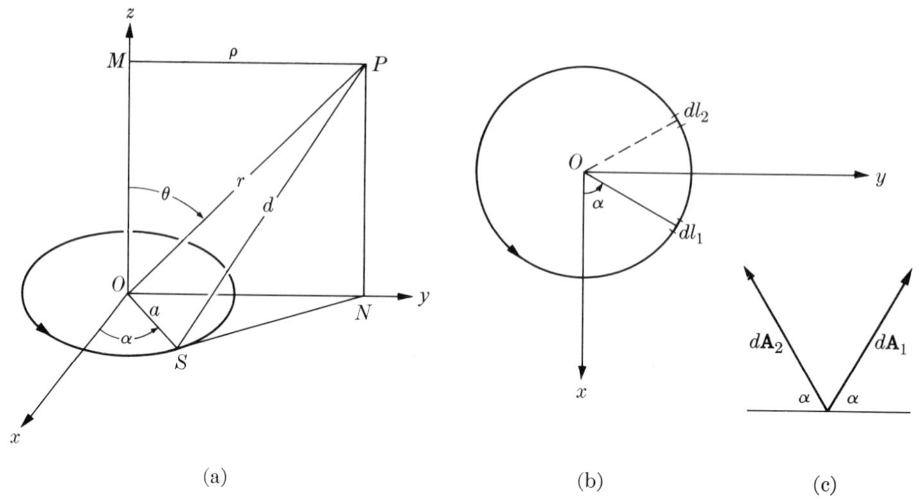

FIG. 6–9. The calculation of the magnetic field of the ring current.

It is also of interest to calculate the field at points off the axis. This calculation is more involved and is done most simply by using the vector potential. Let us consider a current ring as shown in Fig. 6–9(a) and the vector potential at some point P. We see that this problem has symmetry about the axis of the ring, which we will choose to be the z-direction. We drop a perpendicular PN from P to the plane containing the ring (i.e., the xy-plane) and let ON define the y-direction. If we now calculate the field line pattern in the yz-plane, the complete result is obtained by rotating this pattern about the Oz-axis.

From the relation

$$d\mathbf{A} = \frac{\mu I\, d\mathbf{l}}{4\pi r},$$

we see that the contribution to the vector potential from an element of current $d\mathbf{l}$ is in the same direction as $d\mathbf{l}$. Let us consider the two elements $d\mathbf{l}_1$ and $d\mathbf{l}_2$ situated as shown in Fig. 6–9(b). The contributions $d\mathbf{A}_1$ and $d\mathbf{A}_2$ to the vector potential will be equal, since $d\mathbf{l}_1$ and $d\mathbf{l}_2$ are at the same distance from P, and will be inclined to the y-direction as shown in Fig. 6–9(c). The net result is a vector parallel to the plane of the ring and perpendicular to the radius MP (or OP). Note that it is important to refer directions, etc., to the physical aspects of the problem (i.e., the current ring and not the xyz-axes.) This is because we have chosen the axis in a special way in order to simplify later calculations. (For example, the vector potential at P with the axis chosen in this particular way is in the $-x$-direction, but had we chosen to let ON define the x-axis, it would have been in the y-direction.) We therefore see that the vector potential has constant magnitude on a circle centered on the axis of the ring, and its direction is tangential to the circle. We must now calculate this magnitude.

For the element dl at S we have

$$dA = \frac{\mu I\,dl}{4\pi PS} = \frac{\mu I a\,d\alpha}{4\pi d}.$$

We obtain d from the geometry of the problem, and to simplify matters we suppose that the distance from P to the current ring is large compared with the radius of the ring:

$$\begin{aligned} SN^2 &= OS^2 + ON^2 - 2OS \cdot ON \cos \angle SON = a^2 + y^2 - 2ay \sin \alpha, \\ PS^2 &= PN^2 + SN^2, \\ d^2 &= z^2 + y^2 + a^2 - 2ay \sin \alpha = r^2 + a^2 - 2ar \sin \theta \sin \alpha, \end{aligned}$$

and assuming $r \gg a$, we have

$$\frac{1}{d} \simeq \frac{1}{r}\left(1 - \frac{2a}{r} \sin \theta \sin \alpha\right)^{-1/2} \simeq \frac{1}{r}\left(1 + \frac{a}{r} \sin \theta \sin \alpha\right).$$

From Fig. 6–9(c) the tangential component of the contribution to the vector potential is given by $dA \sin \alpha$, and hence the total potential is

$$A = \int_0^{2\pi} \frac{\mu I a \sin \alpha}{4\pi r}\left(1 + \frac{a}{r} \sin \theta \sin \alpha\right) d\alpha,$$

and carrying out the integration, we obtain

$$A = \frac{\mu I a^2 \sin \theta}{4r^2}.$$

It is simplest to work the problem from here in spherical polar coordinates. The direction of the **A**-vector then becomes the $\mathbf{u}_\phi$-direction, and the vector has only one component. To obtain **B** we must calculate curl **A**. The calculation is in principle exactly the same as in Cartesian coordinates, but the calculation is more tedious. It is carried through in the Appendix. For the case where **A** has only a ϕ-component,

$$\begin{aligned} \text{curl}\,\mathbf{A} &= \mathbf{u}_r \frac{1}{r \sin \theta} \frac{\partial}{\partial \theta}(A_\phi \sin \theta) - \mathbf{u}_\theta \frac{1}{r} \frac{\partial}{\partial r}(r A_\phi), \\ \therefore \mathbf{B} &= \frac{\mu I a^2}{4}\left(\mathbf{u}_r \frac{1}{r \sin \theta} \frac{2 \sin \theta \cos \theta}{r^2} + \mathbf{u}_\theta \frac{1}{r} \frac{\sin \theta}{r^2}\right) \\ &= \frac{\mu I a^2}{4}\left(\mathbf{u}_r \frac{2 \cos \theta}{r^3} + \mathbf{u}_\theta \frac{\sin \theta}{r^3}\right). \end{aligned}$$

Noting that πa^2 is the area S of the ring, we can write

$$\mathbf{H} = \frac{IS}{4\pi r^3}(\mathbf{u}_r\, 2 \cos \theta + \mathbf{u}_\theta \sin \theta).$$

Let us compare this with the expression

$$\mathbf{E} = \frac{P}{4\pi\epsilon r^3}(\mathbf{u}_r\, 2\cos\theta + \mathbf{u}_\theta \sin\theta),$$

which we obtained for the electrostatic dipole. The variation with r and θ is seen to be identical. If we take the limit $a \to 0$ so that $S \to O$, and let $I \to \infty$ in such a way that $IS = M$ remains constant, then M corresponds to P, the dipole moment. Thus the current loop produces exactly the same **H**-field variation as the **E**-field variation produced by a charge dipole. The current loop may therefore be regarded as a magnetic dipole of moment M. We can also define **M** as a vector by associating it with the direction given by the right-hand rule applied to the direction of circulation of current around the ring.

6-9 Uniqueness theorem

In the previous sections we calculated fields due to distributions of electric charges and electric currents. We saw that fields may originate either from sources or vortexes. The electric charges are the sources of the electric fields ($\text{div}\,\mathbf{D} = q_v$), and the currents are the vortexes of the magnetic field ($\text{curl}\,\mathbf{H} = \mathbf{J}$). Hence what we have really done is to calculate the field, given its sources or its vortexes. This is equivalent to solving a partial differential equation. We know that when we integrate an ordinary differential equation, we can add an arbitrary constant to the solution, and the resulting function still satisfies the differential equation. We now look a little further into the integration of these equations. Let us first consider the equation $\text{div}\,\mathbf{D} = q_v$. We know that the operation div curl is identically equal to zero. Suppose that we have a function $\mathbf{D}_1$ (a vector point function) which satisfies our original differential equation, namely, $\text{div}\,\mathbf{D}_1 = q_v$. If we add to this curl **F**, where **F** is any arbitrary vector field forming a second point function (or field) $\mathbf{D}_2$, then $\mathbf{D}_2$ also satisfies the equation, since

$$\mathbf{D}_2 = \mathbf{D}_1 + \text{curl}\,\mathbf{F},$$

$$\text{div}\,\mathbf{D}_2 = \text{div}\,\mathbf{D}_1 + \text{div curl}\,\mathbf{F} = \text{div}\,\mathbf{D}_1 = q_v.$$

Similarly we can add to any solution, $\mathbf{H}_1$, of the equation $\text{curl}\,\mathbf{H} = \mathbf{J}$, the field grad ψ, where ψ is an arbitrary scalar point function, since curl grad is identically zero.

In the solution of the ordinary differential equation we remove the arbitrariness of the constant by specifying an initial condition, and the constant can then be determined uniquely. We have shown in a general way that we must specify boundary conditions in solving partial differential equations. We now wish to know exactly what must be specified so that the field can be determined completely, i.e., that it is unique.

If we consider a general vector field, then we have

$$\text{curl } \mathbf{U} = \mathbf{J}, \qquad \text{div } \mathbf{U} = m,$$

where m and $\mathbf{J}$ are the sources and vortexes of $\mathbf{U}$. We want to determine the conditions under which the above equations have only one solution. Let us first suppose that we have the two solutions $\mathbf{U}_1$ and $\mathbf{U}_2$ which are different. The difference field $\delta\mathbf{U} = \mathbf{U}_1 - \mathbf{U}_2$ will then satisfy the equations

$$\text{curl } \delta\mathbf{U} = 0, \qquad \text{div } \delta\mathbf{U} = 0.$$

Since curl $\delta\mathbf{U} = 0$, we can write $\delta\mathbf{U} = \text{grad } \phi$, where ϕ is a scalar field. Hence

$$\nabla^2\phi = 0,$$

and ϕ satisfies Laplace's equation. From the vector identity (Appendix) we have

$$\text{div } (\phi\delta\mathbf{U}) = \delta\mathbf{U} \cdot \text{grad } \phi + \phi \text{ div } \delta\mathbf{U},$$

and since $\delta\mathbf{U} = \text{grad } \phi$ and div grad $\phi = 0$, we have

$$\text{div } (\phi\delta\mathbf{U}) = \text{grad } \phi \cdot \text{grad } \phi = |\text{grad } \phi|^2.$$

By integrating this throughout a volume and applying the divergence theorem, we obtain

$$\iiint |\text{grad } \phi|^2 \, d\tau = \oiint \phi \text{ grad } \phi \cdot d\mathbf{S}.$$

The integral on the right-hand side will be zero if grad $\phi \cdot d\mathbf{S}$ is zero on S. If the normal component of $\mathbf{U}$ is specified on S, then the normal component of $\delta\mathbf{U}$ is zero on S and so $\partial\phi/\partial n$ is zero on S. Since grad $\phi \cdot d\mathbf{S} = \text{grad } \phi \cdot \mathbf{u}_n \, d\mathbf{S} = (\partial\phi/\partial n) \, dS$, then grad $\phi \cdot d\mathbf{S}$ will be zero where $(\partial\phi/\partial n)$ is zero. Hence the integral on the right-hand side will be zero if the normal component of $\mathbf{U}$ is specified on S. If the integral on the left is zero, then grad ϕ must be zero, since $|\text{grad } \phi|^2$ cannot be negative, and the only way in which the sum of nonnegative numbers can be zero is for each of them to be zero individually. If grad ϕ is zero, then $\delta\mathbf{U}$ is zero, $\mathbf{U}_1 = \mathbf{U}_2$, and the field is unique. Thus we can say that the field $\mathbf{U}$ is uniquely defined within the region S by the sources within it (m and $\mathbf{J}$) and the normal component of $\mathbf{U}$ on S.

It may be shown that the field $\mathbf{U}$ is also uniquely specified within a region by the sources in it and the *tangential* component of $\mathbf{U}$ over the surface.

Problems

6–1 Use the result for the potential on the axis of a ring charge to calculate the potential on the axis of a circular disk carrying a surface charge density q_s per unit area.

6–2 Calculate the field for Problem 6–1.

6–3 Calculate the potential due to an infinite straight-line charge density of magnitude q_l per unit length.

6–4 Use the results of Problem 6–3 to obtain the potential due to two parallel infinite straight-line charges of density $+q_l$ and $-q_l$ per unit length separated by a distance d. [*Note:* This forms a twin-wire transmission line with zero radius for the conductors. Approximately the same result will be obtained in practice if the separation of the conductors is large compared to their radius.]

6–5 Calculate the field in Problem 6–4 on a perpendicular line joining the two charges.

6–6 Assume that in Problem 6–4 the charges carried by the wires are ± 1 coul/m, and the wires are separated by 1 m. Calculate the field at a point midway between the wires and at a point on the line joining the two wires distant 1.5 m from the center point.

6–7 Calculate the field for a coaxial transmission line if the inner conductor carries a charge q_l per unit length and the outer carries $-q_l$ per unit length.

6–8 A diode consists of two parallel infinite conducting plates separated by a distance d and with potential V_b between them. The space charge is assumed to consist of an infinite slab of thickness t, one side of the slab coinciding with the cathode. The electron density in the slab is assumed constant. In the remainder of the space between the electrodes the electron density is negligible. Calculate the potential by solving Poisson's equation.

6–9 Calculate the magnetic field for a current flowing parallel to the axis of a cylinder of radius a with uniform density $\mathbf{J}_1$ per unit area inside the cylinder and zero outside it. Consider the field both inside and outside the cylinder.

6–10 Show that the field at the center of a ring current is given by $H = I/2a$.

6–11 Calculate the magnetic field on the axis of the ring of Problem 6–10.

6–12 Calculate the field on the axis of a solenoid of finite length. Use the result of Problem 6–11.

6–13 Calculate the magnetic field in a coaxial cable carrying a steady current.

6–14 Each of two infinitely long parallel wires separated by a distance d carries a current I. Show that the force F on a length l of one of the wires is given by $F = \mu_0 I^2 l/(2\pi d)$.

6–15 Calculate the force per unit length in Problem 6–14, given that the wires are 1 m apart in vacuum and carry a current of 1 amp.

6–16 Calculate the torque on a rectangular loop of wire carrying current I in the presence of a magnetic field in the plane of the loop and perpendicular to one pair of sides of the rectangle. By letting the area of the loop become infinitesimal show that the torque on a magnetic dipole is given by $\mathbf{m} \times \mathbf{B}$.

6–17 By following arguments similar to those used in the proof of the uniqueness theorem, show that if ϕ satisfies Laplace's equation, that is, $\nabla^2\phi = 0$, and if ϕ is constant everywhere outside the surface S, then it is constant everywhere.

CHAPTER 7

The Calculation of Resistance, Capacitance, and Inductance

7–1 The calculation of resistance

Our fundamental definition of resistance is

$$R = \frac{V}{I}.$$

The resistor is a two-terminal network, and the resistance between these two terminals is obtained from the above relation, where V is the voltage measured across the two terminals and I is the total current flowing between them. The resistance will obviously depend on the medium between these two terminals, and we shall now show how it is calculated in terms of the properties of the material between the two terminals and its geometry. As was already noted, a complication arises if the medium is not linear, and thus we shall confine our treatment to linear media.

As a very elementary example, we may take a rectangular slab of material of conductivity σ, with two perfectly conducting plates at its ends forming the two terminals, as shown in Fig. 7–1. From the symmetry of this problem we see that the electric field and the electric current will everywhere be parallel to the long side of the bar, and that they will have the same value at all points. The voltage V is therefore given by

$$V = \int E\,dl = El,$$

and the current I by

$$I = \iint J\,dS = Jab.$$

Since $J = \sigma E$, we have the relation

$$R = \frac{V}{I} = \frac{El}{Jab} = \frac{l}{\sigma\,ab}.$$

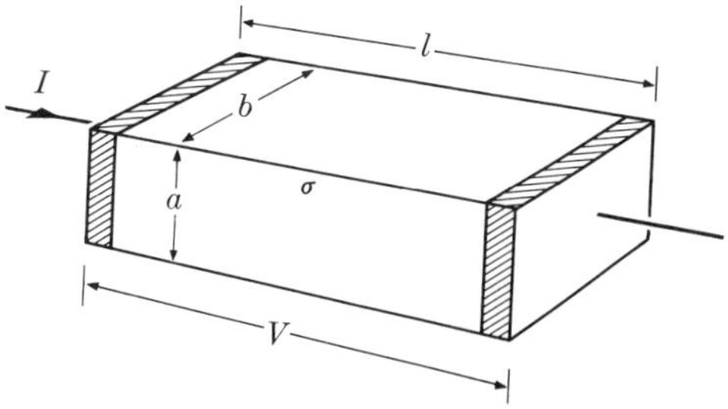

FIG. 7–1. A bar resistor.

As a more complicated example, let us suppose that the resistor of Fig. 7–1 is bent so as to take the form shown in Fig. 7–2. The long edge now becomes the arc of a circle, the other edges remaining straight lines. This problem has a

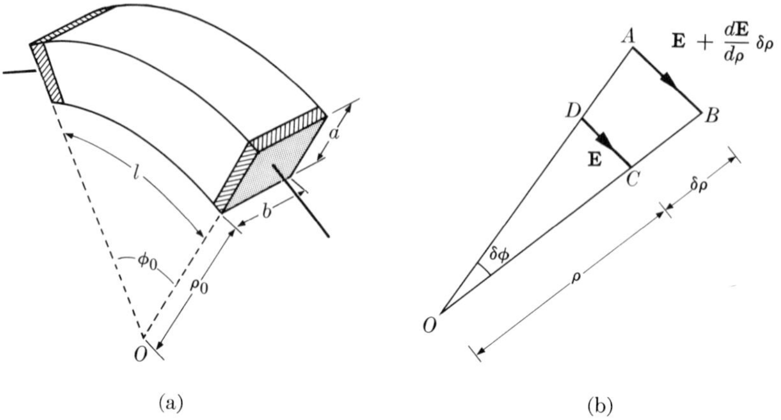

FIG. 7-2. A curved bar resistor.

cylindrical symmetry and has an axis passing through the point O (the center of the circle) parallel to the side of the length b. The equipotential surfaces are the intersections of the resistor and the planes passing through the axis, and the current flow lines are circles centered on the axis. We may suppose from symmetry that the **E**-vector has the same value at all points which are at the same distance from the axis, but will not have the same value at all points across the cross section of the resistor. To determine how it varies we use the relation curl $\mathbf{E} = 0$. Let us consider the element of area $ABCD$ shown in Fig. 7-2(b). By taking the line integral $\oint \mathbf{E} \cdot d\mathbf{l}$ around $ABCD$, we obtain

$$\oint \mathbf{E} \cdot d\mathbf{l} = \left(E + \frac{dE}{d\rho}\,\delta\rho\right)(\rho + \delta\rho)\delta\phi - E\rho\,\delta\phi,$$

the first term being the contribution from AB and the second from CD. The above expression must be zero since curl **E** is zero, and hence

$$E\,\delta\rho + \rho\frac{dE}{d\rho}\,\delta\rho = 0, \qquad \text{or} \qquad \frac{dE}{E} = -\frac{d\rho}{\rho},$$

$$\therefore \log E = -\log\rho + \text{const},$$

and

$$E = \frac{k}{\rho}.$$

Furthermore,

$$I = \iint \mathbf{J} \cdot d\mathbf{S} = \iint \sigma\mathbf{E} \cdot d\mathbf{S} = b\sigma\int_{\rho_0}^{\rho_0+a} \frac{k}{\rho}\,d\rho = kb\sigma\log\left(1 + \frac{a}{\rho_0}\right).$$

The voltage is given by

$$V = \int \mathbf{E} \cdot d\mathbf{l} = \int \frac{k}{\rho}\, \rho\, d\phi = k\phi_0 = \frac{kl}{\rho_0}.$$

Hence for the resistance we have

$$R = \frac{V}{I} = \frac{1}{\sigma} \frac{l}{\rho_0 b \log(1 + a/\rho_0)}.$$

7–2 The calculation of capacitance

The fundamental definition of capacitance is

$$C = \frac{Q}{V}.$$

The capacitor is a two-terminal network, and the capacitance between these two terminals is obtained from the above expression, where V is the voltage measured across the terminals and Q is the total charge supplied via the terminals. Like resistance, capacitance depends on the properties of the medium between the two terminals and on the geometrical configuration of the material. Again we confine our discussion to linear media.

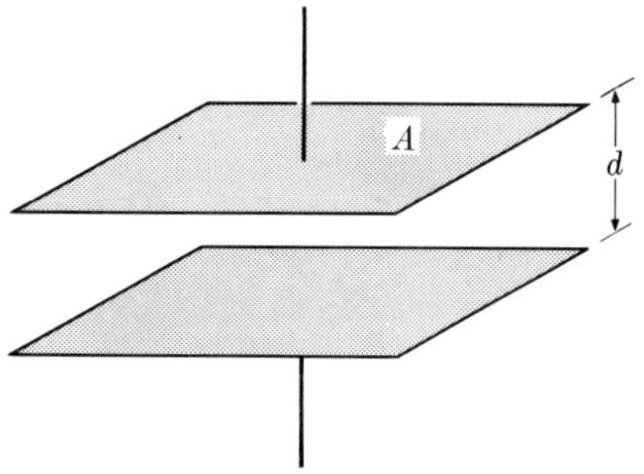

FIG. 7–3. The parallel-plate capacitor.

As a simple example of capacitance, let us consider the perfectly conducting parallel rectangular plates shown in Fig. 7–3. We shall suppose that $\mathbf{E}$ and $\mathbf{D}$ are uniform and everywhere perpendicular to the plates. If the charge on the plate is Q, then by Gauss' theorem we have $D = Q/A$ and $V = \int \mathbf{E} \cdot d\mathbf{l} = Ed$. The properties of the medium between the plates give the relation between $\mathbf{D}$ and $\mathbf{E}$, namely, $\mathbf{D} = \epsilon\mathbf{E}$, and hence

$$C = \frac{Q}{V} = \frac{DA}{El} = \frac{\epsilon A}{l}.$$

Our assumption that the field lines are everywhere perpendicular to the plates is not strictly correct. There are fringe effects occurring at the edge of the

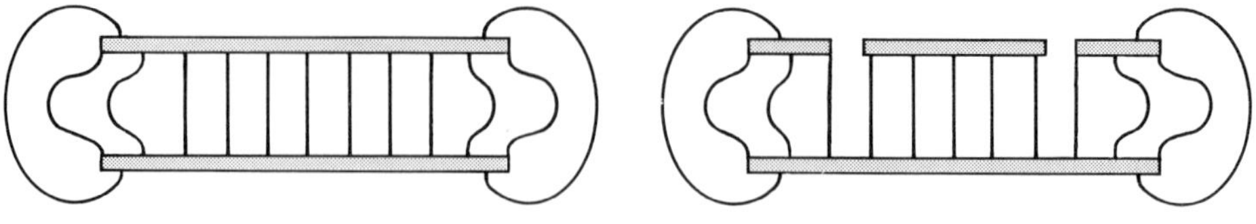

FIG. 7-4. (a) The field of a parallel-plate capacitor. (b) The effect of a guard ring.

plates, as shown in Fig. 7-4(a). These are difficult to take into account in the calculation. In experiments where it is essential to calculate the capacity in terms of the geometry of the capacitor, a guard-ring capacitor is used. This has one plate which is smaller than the other, and the small plate is surrounded by a guard ring as shown in Fig. 7-4(b). The guard ring and the upper plate are always kept at the same voltage, but the guard ring does not form part of the capacitance. This produces a field pattern as shown, and our assumptions regarding the field are now strictly fulfilled. The above formula using the area of the smaller plate A now gives an accurate value for the capacitance.

Let us consider next a coaxial cable which we suppose to be straight and of infinite length. We are thus determining the capacity between two concentric cylinders of infinite length. The symmetry of the problem shows that the **D**- and **E**-lines will be radial lines passing from one cylinder to the other. Since there are no charges in the region between the cylinders, div $\mathbf{D} = 0$. Let us consider a small volume of axial length l and cross section $ABCD$, as shown in Fig. 7-5. The only flux is over the curved surfaces, and hence for div **D** to be zero we must have

$$Dl\rho\ \delta\phi + \left(D + \frac{dD}{d\rho}\,\delta\rho\right) l(\rho + \delta\rho)\ \delta\phi = 0,$$

from which

$$\frac{dD}{D} = -\frac{d\rho}{\rho}, \qquad \text{and} \qquad D = \frac{k}{\rho}.$$

If the charge per unit length of this cylinder is Q_l, then by Gauss' theorem applied to a cylindrical surface of length l and of radius infinitesimally larger than a, we obtain

$$2\pi a\, D_a = Q_l l,$$

$$\therefore \frac{Q_l l}{2\pi a} = \frac{k}{a},$$

and

$$k = \frac{Q_l l}{2\pi},$$

$$\therefore D = \frac{Q_l l}{2\pi\rho}.$$

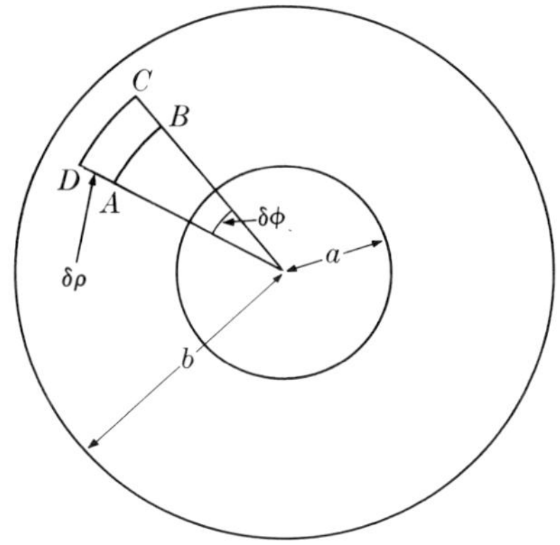

FIG. 7–5. A cross section of a coaxial cable.

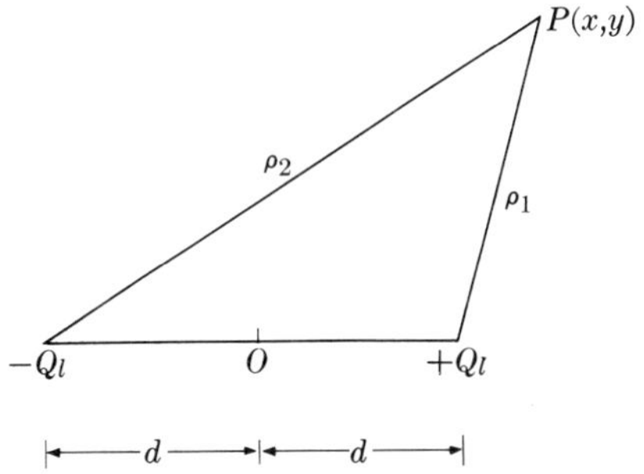

FIG. 7–6. Two infinitely long parallel-line charges.

From the relation $D = \epsilon E$, it then follows that

$$E = \frac{Q_l l}{2\pi\epsilon\rho}, \qquad \text{and} \qquad V = \int_a^b \frac{Q_l l}{2\pi\epsilon\rho}\, d\rho = \frac{Q_l l}{2\pi\epsilon} \log\frac{b}{a},$$

and hence

$$C = \frac{Q_l l}{V} = \frac{2\pi\epsilon l}{\log(b/a)}.$$

The capacity per unit length, the parameter needed in transmission line theory, is thus

$$C_l = \frac{2\pi\epsilon}{\log(b/a)}.$$

The above examples illustrate that the problem of calculating capacitance is essentially the same as the problem of finding the field and integrating to get the voltage. Alternatively it may be more convenient in some cases to work with equipotential surfaces. The examples considered above are straightforward as far as the calculations are concerned, and the fields were obtained mainly by the use of the symmetry properties of the problem. An extensive study has been made of the calculation of fields in more complicated cases, and various mathematical techniques have been formulated. As an example of a somewhat more difficult case, we take the parallel-wire transmission line.

Let us first consider two infinitely long line charges (Fig. 7–6). It may be shown (see Problem 6–3) that the potential V at a point distant ρ from an infinite line charge of Q_1 coul/m is given by

$$V - V_0 = -\frac{Q_l}{2\pi\epsilon} \log \rho,$$

and if we define the zero of voltage $V = 0$ for $\rho = 1$, then

$$V = \frac{Q_l}{2\pi\epsilon} \log \frac{1}{\rho}.$$

Hence the potential at P in Fig. 7-6 is

$$\frac{Q_l}{2\pi\epsilon} (\log \rho_2 - \log \rho_1) = \frac{Q_l}{4\pi\epsilon} \log \frac{\rho_2}{\rho_1}.$$

Expressing this in terms of the xy-coordinates of P and the separation between the lines, $2d$, we have

$$\frac{Q_l}{2\pi\epsilon} \frac{1}{2} \log \left[\frac{y^2 + (x - d)^2}{y^2 + (x + d)^2} \right].$$

Hence, for an equipotential surface, we obtain

$$\frac{Q_l}{2\pi\epsilon} \frac{1}{2} \log \frac{y^2 + (x - d)^2}{y^2 + (x + d)^2} = V,$$

or

$$\log \frac{y^2 + (x - d)^2}{y^2 + (x + d)^2} = k,$$

where $k = 2V(2\pi\epsilon/Q_l)$. On rearranging, we have

$$y^2 + \left(x - \frac{e^{k/2} + e^{-k/2}}{e^{k/2} - e^{-k/2}} d \right)^2 = \left(\frac{2}{e^{k/2} - e^{-k/2}} \right)^2 d^2,$$

or

$$y^2 + (x - d \coth \tfrac{1}{2}k)^2 = (\operatorname{cosech} \tfrac{1}{2}k)^2 d^2.$$

This is the equation of a circle whose center is on the x-axis at $x = d \coth \frac{1}{2}k$ and of radius $d \operatorname{cosech} \frac{1}{2}k$. If we choose an equipotential surface of potential V, then it has radius $a = d \operatorname{cosech} (2\pi\epsilon V/Q_l)$ and its center is at $x = d \coth (2\pi\epsilon V/Q_l)$. Similarly the surface of potential $-V$ has the same radius, and its center is at $x = -d \coth (2\pi\epsilon V/Q_l)$. These surfaces are shown in Fig. 7-7. Since

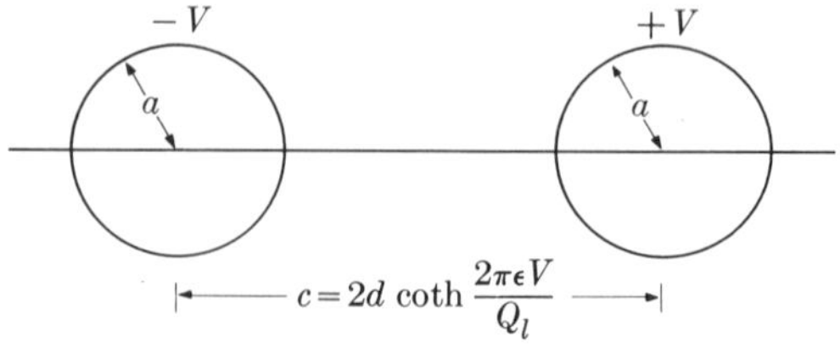

FIG. 7-7. The parallel-wire transmission line.

they are equipotentials, they can be replaced by conductors. We would then have two infinitely long cylindrical conductors of radius a carrying charges per unit length of $+Q_l$ and $-Q_l$ and with a distance between their centers equal to c. This is the case for the parallel-wire transmission line. We know further that the voltage between the conductors is given by $2V$. Thus we have the equations

$$a = d \operatorname{cosech} \frac{2\pi\epsilon V}{Q_l}, \qquad c = 2d \coth \frac{2\pi\epsilon V}{Q_l},$$

and hence

$$\frac{c}{2a} = \cosh \frac{2\pi\epsilon V}{Q_l}.$$

Since the voltage between the two wires is $2V$, the capacitance per unit length is given by

$$C_l = \frac{Q_l}{2V} = \frac{\pi\epsilon}{\cosh^{-1}(c/2a)},$$

and if $c \gg a$, this may be written as

$$C_l = \frac{\pi\epsilon}{\log(c/a)}.$$

Note that this problem was solved by a rather indirect approach. We calculated the capacitance per unit length of the parallel-wire transmission line by using results obtained from the investigation of the field due to infinite line charges. This is an example of how the results of a problem which at first sight may be of academic interest only can have great practical use—a not infrequent occurrence.

7–3 The calculation of inductance

Proceeding in an analogous manner, we begin with our fundamental definition of inductance,

$$L = \frac{\psi}{I}.$$

The inductor is a two-terminal network, and I in the above formula is the current flowing between the two terminals. However, ψ does not lend itself to quite so straightforward an interpretation as all the other quantities with which we have been concerned in calculating resistance and capacitance. It is to be interpreted as the total flux linkage in the circuit. The inductance also is dependent on the properties and the geometrical arrangement of the medium between the two terminals, and, as before, we shall still confine our discussion to linear media.

As a simple example of the calculation of inductance, let us consider the case of the coaxial cable (Fig. 7-8). The **H**-field is readily obtained by Ampère's theorem. By symmetry the **H**-lines will be concentric circles. A circular path of integration of radius greater than b encircles zero current, since the two conductors carry equal and opposite currents. Hence the field outside the cable is zero. We suppose that the radius of the inner conductor is small compared to that of the outer, so that the flux inside the inner conductor can be neglected. The flux linked with the current I is then given by the integral of the flux density across the region between the two conductors. For a radius ρ we have

$$2\pi\rho H = I, \qquad \text{or} \qquad H = \frac{I}{2\pi\rho}\cdot$$

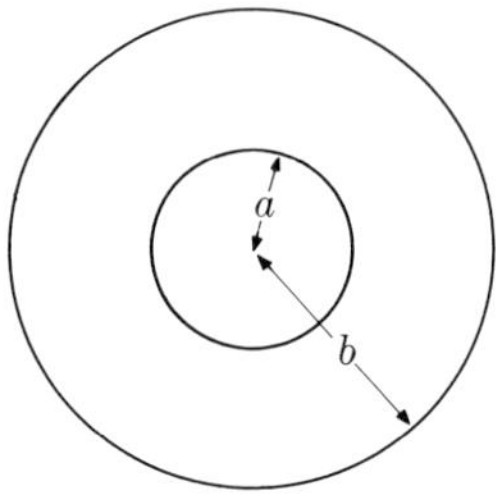

FIG. 7-8. The coaxial cable.

Considering the length l of the cable, we find that the flux is given by

$$\psi = \int_a^b \frac{\mu I l}{2\pi\rho}\,d\rho = \frac{\mu I l}{2\pi}\log\frac{b}{a},$$

and hence

$$L = \frac{\mu l}{2\pi}\log\frac{b}{a}\cdot$$

The inductance per unit length of the cable, the parameter needed in transmission line theory, is

$$L_l = \frac{\mu}{2\pi}\log\frac{b}{a}\cdot$$

We can also obtain the inductance per unit length of the parallel-wire line (Fig. 7-9). The field at the point P consists of two components, one due to the current $+I$ in the left-hand wire and one due to the current $-I$ in the right-hand wire. Both components are perpendicular to the line joining the center of the two wires, and the resultant field is

$$H = \frac{I}{2\pi}\left(\frac{1}{x} + \frac{1}{c - x}\right) = \frac{I}{2\pi}\frac{c}{x(c - x)}\cdot$$

If we make the approximation $c \gg a$, then we can neglect the flux inside the

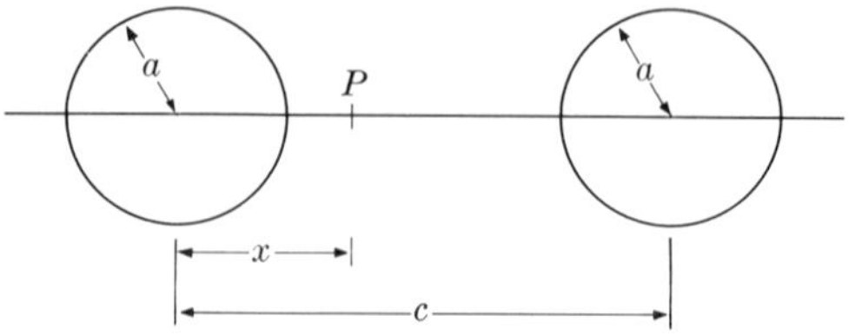

FIG. 7-9. The parallel-wire transmission line.

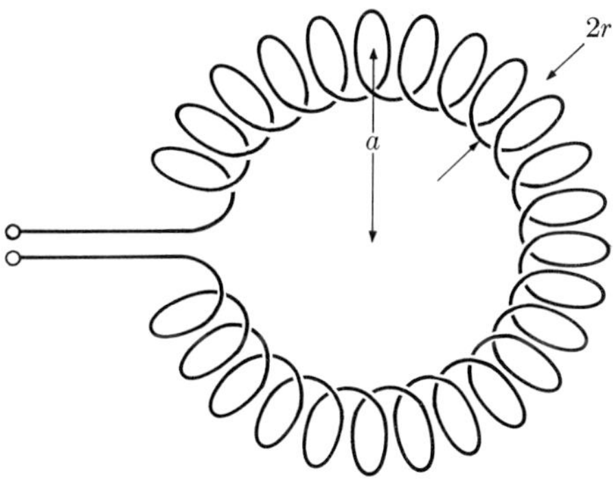

FIG. 7-10.

wire. The flux linked with the circuit is given by the integral of the flux density over the region between the two wires. Hence, considering unit length, we have

$$\psi_l = \frac{\mu I}{2\pi}\int_a^{c-a}\left(\frac{1}{x} - \frac{1}{c-x}\right)dx$$

$$= \frac{\mu I}{2\pi}\left(\log\frac{c-a}{a} - \log\frac{a}{c-a}\right) = \frac{\mu I}{2\pi}\log\left(\frac{c-a}{a}\right)^2,$$

and since $c \gg a$,

$$\psi_l = \frac{\mu I}{\pi}\log\frac{c}{a}\cdot$$

The inductance per unit length is then given by

$$L_l = \frac{\mu}{\pi}\log\frac{c}{a}\cdot$$

As an example of a lumped circuit element we take a toroidal coil (Fig. 7-10). Let us suppose that the radius of the toroid a is large in comparison with the radius of the cross section r. By symmetry the field lines will be circles. We can see by Ampère's theorem that the magnetic flux must be confined to the inside of the coil, since any circular contour lying entirely outside the coil will enclose no current, and therefore the field must be zero. This argument will be true for all circular contours of radius greater than $a + r$ or less than $a - r$. Since $a \gg r$, for a circular contour inside the coil we have

$$2\pi a H = nI,$$

where n is the total number of turns in the coil. We could obtain the same result from the boundary condition for H. If the total number of turns is n, then the number per unit length is $n/2\pi a$, and hence the current per unit length is $nI/2\pi a$, which is the current sheet J_s. Since the field outside the current sheet is zero, the field inside is equal to the discontinuity in the **H**-field, which is equal to J_s. The flux is therefore given by $\mu A n I/2\pi a$, where A is the cross-sectional area of the coil. Note, however, that this flux is linked with the circuit n times, since each flux line is linked with all n turns of the circuit, each of which carries the current I. Hence the total flux linkage ψ is given by $\mu A n^2 I/2\pi a$, and thus for the inductance we have

$$L = \frac{\mu A n^2}{2\pi a}.$$

The inductance of a long solenoid can be obtained by similar reasoning. If the solenoid is long compared with its radius, we can consider that all of the flux passes through the whole solenoid, and if the total number of turns in the solenoid is n and its length is l, then the inductance is given by

$$L = \frac{\mu A n^2}{l}.$$

However, the assumption that all of the flux passes through the whole solenoid is only approximate. In practice, the field lines will be as shown in Fig. 7–11. Some of the flux lines "leak out" of the coil and are not linked with all the turns. This leakage flux is difficult to allow for in theoretical calculations.

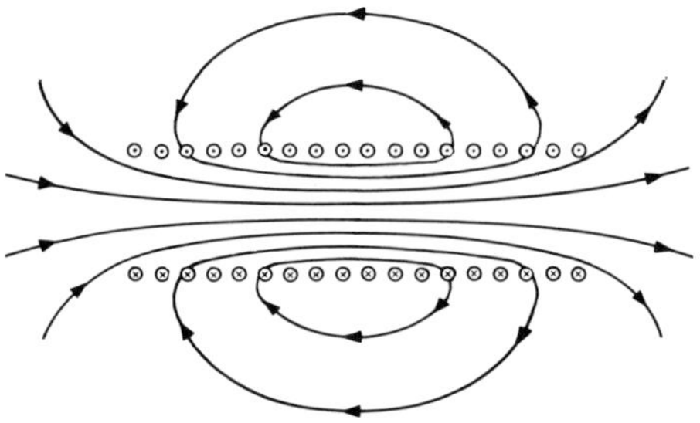

FIG. 7–11. A cross section of a solenoid and its magnetic field.

The above ideas can be extended to the calculation of mutual inductance. If we wind a coil of n' turns on top of the toroidal coil, then the flux due to the current I flowing in the original coil linked with the second coil is given by $\mu A n n' I/2\pi a$, since now each flux line is linked with n' turns of the second coil. The mutual inductance is therefore

$$M = \frac{\mu A n n'}{2\pi a}.$$

Similarly, if a short coil is wound near the center of a long solenoid so that, to a good approximation, all of the flux due to the solenoid is linked with all n' turns of the second coil, then

$$M = \frac{\mu A n n'}{l}.$$

Note how much more difficult it would be to calculate M in the opposite way, i.e., to pass a current I through the second coil and calculate the flux due to the current linked with the first. However, we have already proved that the same result must be obtained whichever way the calculation is made.

Problems

7–1 Calculate the resistance of a transformer core consisting of a toroid with a small gap Δ in it. Assume the radius of the toroid to be large compared with the radius of the cross section.

7–2 Calculate the capacity between two rectangular plates of side a and b if the sides of length a are parallel, one pair being distance d apart and the other $d + \alpha$. Show that your result reduces to that of the parallel-plate capacitor when $\alpha = 0$. [*Hint:* Assume that the field lines are circular arcs.]

7–3 Use the results of Problem 7–2 to obtain the capacity of a parallel-plate capacitor with plates 10 cm square and separated by 1 cm. Calculate the capacity, given that the capacitor is not quite true mechanically and $\alpha = 1$ mm.

7–4 Calculate the capacity of a parallel-plate capacitor whose plates have become slightly curved in one direction. The edge of length a is straight, and the edge of length b has a radius of curvature R for the inner plate and $R + t$ for the outer, where t is the separation between the plates.

7–5 Calculate the capacity of a capacitor consisting of two concentric spheres.

7–6 Calculate the inductance of a solenoid of length 10 cm and radius 1 cm consisting of 100 turns.

7–7 An infinitely long parallel-plate transmission line consists of two infinite planes separated by a distance d. Calculate the inductance of a length l and width b of this line.

7–8 An infinitely long straight wire of radius a carries a uniform current density. The return current is carried by a tube with walls of negligible thickness and radius a, the tube being insulated from the wire. Calculate the inductance of a length l of this configuration.

7–9 Calculate the inductance per unit length for a coaxial cable whose inner wire has a radius of 1 mm and whose outer a radius of 1 cm.

7–10 If a coil of 10 turns is wound on the solenoid (at its center) of Problem 7–6, what is the mutual inductance?

7–11 Two concentric coplanar circular conductors have radii R and r, where $R \gg r$. Calculate the mutual inductance.

7–12 Two coplanar circular conductors have radius r, and their centers are separated by a distance l, where $l \gg r$. Calculate the mutual inductance. [*Hint:* Use the dipole approximation.]

7–13 Calculate the value of the mutual inductance of Problem 7–11, if $R = 25$ cm and $r = 1$ cm.

7–14 Calculate the mutual inductance between an infinitely long straight wire and a square loop, two sides of the loop and the long wire being parallel, and the loop and the wire being in the same plane.

7–15 Assume that the resistance of the curved resistor considered in the text is produced by taking a straight bar of compressible material and bending it to the shape shown, and that the conductivity is proportional to the density. Calculate the resistance.

CHAPTER 8

Material Media

8-1 Introduction

So far we have taken account of the medium in which the fields exist, by introducing the parameters conductivity, permittivity, and permeability in the constitutive relations. In this chapter we consider the last two in somewhat more detail; the permittivity (the electric case) is considered first, and the permeability (the magnetic case) is then treated by analogy as we have done before. This leads to the introduction of two important new vectors, the electric polarization and the magnetization, which are closely connected with the atomic properties of the materials. However, a detailed treatment of the electric and magnetic properties of matter would be out of place here, and this is therefore not pursued very far. In both cases, we first present an elementary argument which illustrates the physical significance of the new vectors. This is followed by a more general treatment involving greater mathematical complexity. In addition to increasing our general understanding of the subject, we shall find the point of view developed here very useful in the treatment of anisotropic media, some simple examples of which will be discussed later.

8-2 Electric polarization

Let us again consider the parallel-plate capacitor discussed in Chapter 3. The relations between the various quantities are

$$E = \frac{V}{l}, \qquad D = \frac{Q}{A}, \qquad D = \epsilon E.$$

We begin by supposing that the space between the plates is a vacuum; then

$$E = \frac{V}{l}, \qquad D_0 = \frac{Q_0}{A}, \qquad D_0 = \epsilon_0 E.$$

Keeping the voltage V constant, we introduce a material medium between the plates with permittivity ϵ. We then have

$$E = \frac{V}{l}, \qquad D = \frac{Q}{A}, \qquad D = \epsilon E.$$

The charge on the plates is now increased by an amount given by

$$Q - Q_0 = A(D - D_0) = AE(\epsilon - \epsilon_0),$$

The increased charge per unit area is therefore $E(\epsilon - \epsilon_0)$.

Let us consider a small volume of the material as shown in Fig. 8–1. We suppose that the ends of the volume have an area ΔA and carry a surface charge density of $+q_s$ at one end and $-q_s$ at the other. We would then have two equal and opposite charges of magnitude $q_s \, \Delta A$ separated by a distance Δl. This constitutes a dipole moment $q_s \, \Delta A \, \Delta l$. But $\Delta A \, \Delta l$ is the volume of the element, and thus the dipole moment per unit volume is q_s.

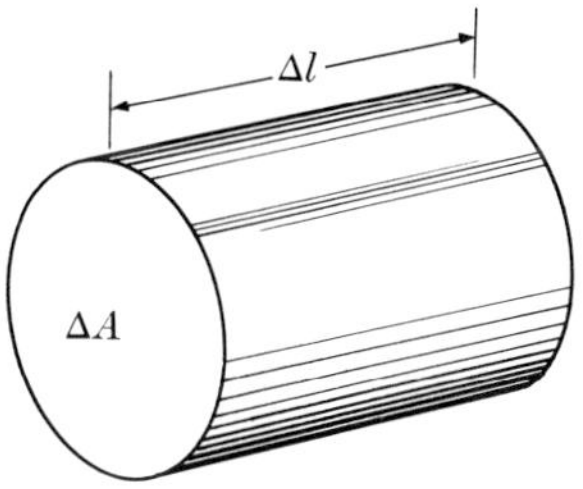

FIG. 8–1. An element of volume.

The atomic theory of matter gives us a simple way of visualizing how this occurs. The atom consists of a heavy positive nucleus surrounded by negative electrons. In a dielectric the electrons are bound to the nuclei and are not free to move throughout the body of the material. However, when an electric field is applied, the positively charged nucleus tends to move in one direction and the electrons in the opposite direction. Although there can be no continuous motion of charges in a dielectric when an electric field is applied, nevertheless the positive and negative charges in it are displaced relative to each other. Thus we can think of each atom acquiring an electric dipole moment. The atoms comprising the volume element above provide the dipole moment per unit volume and thereby produce a surface charge at the surface of the medium. This surface charge can be identified with the change in charge which occurs when the dielectric slab is introduced into the condenser, and hence

$$q_s = E(\epsilon - \epsilon_0).$$

A medium which has developed a dipole moment per unit volume as a result of an applied electric field is said to be *polarized*, and the dipole moment per unit volume is the *polarization*. This is a vector, and in the simplest cases the charges will be displaced in the direction of the field. The polarization vector P is therefore in the same direction as $\mathbf{E}$ and has magnitude q_s. Hence

$$\mathbf{P} = \mathbf{E}(\epsilon - \epsilon_0), \qquad \text{or} \qquad \mathbf{D} = \epsilon_0 \mathbf{E} + \mathbf{P}.$$

Since ϵ_0, the permittivity of a vacuum, is a scalar quantity, $\mathbf{D}$ and $\mathbf{E}$ in a vacuum are in the same direction. Hence we see that if the medium is such that polarization $\mathbf{P}$ is in the direction of the applied field, then $\mathbf{E}$ and $\mathbf{D}$ will be in the same direction in the medium. Taking the magnitudes in the above equation and dividing through by $\epsilon_0 E$ we obtain

$$\kappa = 1 + \chi,$$

where $\kappa = \epsilon/\epsilon_0$ is the dielectric constant, and $\chi = P/\epsilon_0 E$ is the electric susceptibility.

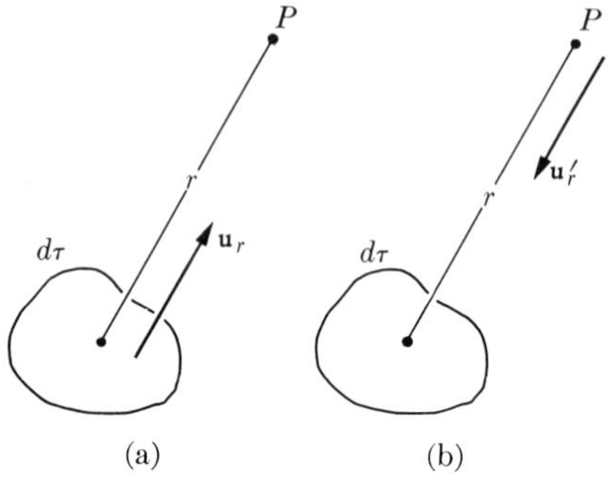

FIG. 8–2. The transfer of the origin from (a) the volume element to (b) the field point.

It is instructive to derive the above result somewhat more generally. If the polarization of the medium is $\mathbf{P}$, then the dipole moment of an element of volume is $\mathbf{P}\,d\tau$. If we consider a point P at a distance r from $d\tau$, as shown in Fig. 8–2(a), then, as we saw in Section 6–5, the voltage at P due to the dipole moment is

$$dV = \frac{\mathbf{P} \cdot \mathbf{u}_r\, d\tau}{4\pi\epsilon r^2},$$

where $\mathbf{u}_r$ is the unit vector drawn from $d\tau$ to P. This may also be written as follows:

$$dV = -\frac{\mathbf{P} \cdot \mathbf{u}_r'\, d\tau}{4\pi\epsilon r^2},$$

where $\mathbf{u}_r'$ is the unit vector drawn from P to $d\tau$. This maneuver merely transfers the origin of our coordinate system from $d\tau$ to P. Let us consider the point P_1 as the origin of the coordinate system. The vector $\mathbf{r}' = r\mathbf{u}_r'$ defines the position of the dipole element with respect to the point at which we are calculating the field. This form is needed here since, if we let $\mathbf{r}'$ vary, we shall move through the dipole moment distribution, and an integral calculated in this manner will give the sum of the effects of all the dipole moments throughout the medium. The expression above may be written as

$$dV = \frac{1}{4\pi\epsilon} \mathbf{P} \cdot \operatorname{grad} \frac{1}{r}\, d\tau.$$

We wish to produce a model in which the effect of the medium is accounted for by the dipole moment distribution; i.e., we replace the medium with permittivity ϵ by a dipole moment distribution situated in vacuum. Consider

$$dV = \frac{1}{4\pi\epsilon_0} \mathbf{P} \cdot \operatorname{grad} \frac{1}{r}\, d\tau.$$

From the vector identity (Appendix A–3),

$$\operatorname{div} w\mathbf{A} = w \operatorname{div} \mathbf{A} + \mathbf{A} \cdot \operatorname{grad} w,$$

we have

$$dV = \frac{1}{4\pi\epsilon_0}\left(\operatorname{div}\frac{\mathbf{P}}{r} - \frac{1}{r}\operatorname{div}\mathbf{P}\right) d\tau,$$

and

$$V = \frac{1}{4\pi\epsilon_0}\iiint \operatorname{div}\frac{\mathbf{P}}{r}\, d\tau - \frac{1}{4\pi\epsilon_0}\iiint \frac{1}{r}\operatorname{div}\mathbf{P}\, d\tau.$$

Transforming the first integral by the divergence theorem, we obtain

$$V = \frac{1}{4\pi\epsilon_0}\iint \frac{\mathbf{P}}{r}\cdot d\mathbf{S} - \frac{1}{4\pi\epsilon_0}\iiint \frac{1}{r}\operatorname{div}\mathbf{P}\, d\tau.$$

Since $\mathbf{P} \cdot d\mathbf{S} = P_n\, dS$, where P_n is the component of P normal to the surface, this represents a surface charge $q'_s\, dS$ on the surface dS. If we let $\operatorname{div}\mathbf{P} = -q'_v$, then

$$V = \frac{1}{4\pi\epsilon_0}\iint \frac{q'_s\, dS}{r} + \frac{1}{4\pi\epsilon_0}\iiint \frac{q'_v\, d\tau}{r}.$$

From Section 6–4 we see that this equation represents the potential due to a volume distribution of charge q'_v and a surface distribution q'_s. In vacuum, $\mathbf{D}_0 = \epsilon_0\mathbf{E}$, and if there is a charge distribution q_v, then

$$\operatorname{div}\mathbf{D}_0 = \operatorname{div}\epsilon_0\mathbf{E} = q_v.$$

In the presence of our dipole moment distribution we have

$$\operatorname{div}\epsilon_0 E = q_v + q'_v = q_v - \operatorname{div}\mathbf{P},$$

and hence

$$\operatorname{div}(\epsilon_0\mathbf{E} + \mathbf{P}) = q_v.$$

In the presence of a medium with permittivity ϵ,

$$\operatorname{div}\mathbf{D} = \operatorname{div}\epsilon\mathbf{E} = q_v.$$

Thus the effect of a medium of permittivity ϵ is the same as that of a dipole moment per unit volume distribution $\mathbf{P}$ given by

$$\mathbf{D} = \epsilon\mathbf{E} = \epsilon_0\mathbf{E} + \mathbf{P}.$$

We have so far neglected the first integral containing the surface charge distribution. This occurs at the boundary, i.e., the surface separating the dipole moment distribution from free space or, more generally, a different medium. If there is a charge distribution q_s on the surface, then the boundary condition (Section 3-7) in vacuum is

$$(D_{0n})_1 - (D_{0n})_2 = \epsilon_0 E_{1n} - \epsilon_0 E_{2n} = q_s.$$

In the presence of our dipole moment distribution there is an additional surface charge q_s', and we have

$$\epsilon_0 E_{1n} - \epsilon_0 E_{2n} = q_s + q_s' = q_s + P_n,$$

$$\therefore \epsilon_0 E_{1n} - (\epsilon_0 E_{2n} + P_n) = q_s.$$

Note that a vector P_n in the same direction as D_n in region 2 (Fig. 8-3) results in a positive surface charge. A polarization vector in region 1 in the same direction as the D-vector in region 1 would produce a negative charge on the surface. If medium 2 has permittivity ϵ, then the boundary condition becomes

$$D_{1n} - D_{2n} = q_s.$$

This result could be extended to the case in which medium 1 has permittivity ϵ_1 and medium 2 permittivity ϵ_2, and we would have

$$(\epsilon_0 E_{1n} + P_{1n}) - (\epsilon_0 E_{2n} + P_{2n}) = q_s.$$

Thus we see that the vector $\epsilon_0 \mathbf{E} + \mathbf{P}$ obeys the same boundary condition as $\mathbf{D}$.

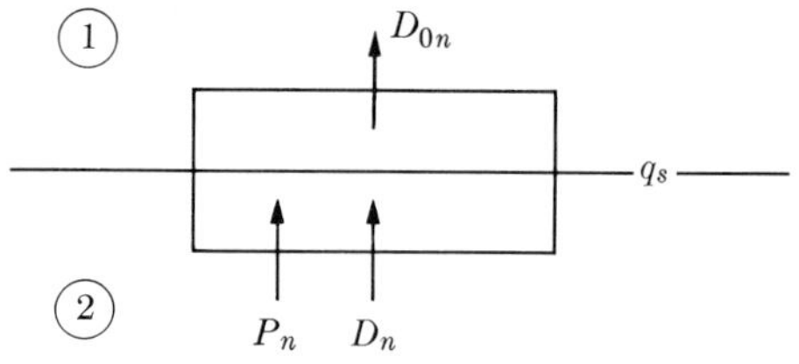

FIG. 8-3. The effect of the surface-charge distribution.

It is more difficult to see the significance of this second derivation. We have shown that if we have a material body with permittivity, situated in an electric field $\mathbf{E}$, then the resulting field is unchanged if the body is replaced by a dipole moment density distribution $\mathbf{P}$ given by $\epsilon \mathbf{E} = \epsilon_0 \mathbf{E} + \mathbf{P}$ situated in vacuum. We did this by first demonstrating that the distribution was equivalent to a volume charge distribution throughout the region formerly occupied by the body, plus a surface charge distribution. If the original problem contained a volume distribution of charge and a surface distribution of charge on the body, in addition to the body itself, then we would have div $\mathbf{D} = q_v$ and, at the boundary, $D_{1n} - D_{2n} = q_s$. We then showed that the vector $\epsilon_0 \mathbf{E} + \mathbf{P}$ obeyed the same equations. Thus we can deal with the effect of matter on our fields either by considering its permittivity ϵ or by considering its polarization $\mathbf{P}$.

The above ideas are put to practical use in calculating the high-frequency dielectric constant of a plasma. The plasma is an ionized gas, and we shall make the following simplifying assumptions. We suppose the medium to contain heavy positive ions and free electrons. The mass of the ions is large enough so that we may assume that they are stationary, and only the electrons respond to the applied field. We suppose that all electrons move independently of each other, i.e., that the motion of one electron does not affect the motion of any of the others, and we neglect collisions between the electrons and the molecules (or ions) of the gas. If x is the displacement of the electron and m is its mass, then the force acting on it is $m\ d^2x/dt^2$, by Newton's law. If this force arises from an applied electric field, then it is equal to eE, where e is the charge on the electron and E is the applied field. Since we have assumed that the electrons are free, this is the only force acting on them, and we have

$$m \frac{d^2x}{dt^2} = eE.$$

To obtain the high-frequency dielectric constant we let $E = E_0 \cos \omega t$, and thus

$$m \frac{d^2x}{dt^2} = eE_0 \cos \omega t.$$

Integrating this twice, we get

$$x = -\frac{e}{m\omega^2} E_0 \cos \omega t = -\frac{eE}{m\omega^2},$$

the constants of integration being made zero by suitable choice of initial conditions. If the electron is displaced a distance x from its equilibrium position, then a dipole moment ex is produced, and if the electron density is N, then we have a dipole moment per unit volume which is

$$P = Nex = -\frac{Ne^2}{m\omega^2} E,$$

and is the polarization of the medium. From this we obtain the susceptibility

$$\chi = \frac{P}{\epsilon_0 E} = -\frac{Ne^2}{\epsilon_0 m\omega^2},$$

and hence the dielectric constant

$$\kappa = 1 - \frac{Ne^2}{\epsilon_0 m\omega^2}.$$

Note that this is independent of the sign of the charge on the particle considered, and also that it is less than one.

8-3 Magnetization

We proceed in an analogous manner for the case of magnetic materials. Let us consider the inductor of Chapter 3, the relations for which are

$$H = \frac{I}{l}, \qquad B = \frac{\psi}{A}, \qquad B = \mu H.$$

If the medium is a vacuum, then

$$H_0 = \frac{I_0}{l}, \qquad B = \frac{\psi}{A}, \qquad B = \mu_0 H_0,$$

and if we specify a constant flux ψ (which would again ensure that no voltage changes appear across the element) and introduce a medium with permeability μ, then we have

$$H = \frac{I}{l}, \qquad B = \frac{\psi}{A}, \qquad B = \mu H.$$

If B is to remain constant and μ is increased, then H must decrease, which implies a decrease in I, and this change in current is given by the expression

$$\frac{I_0 - I}{l} = H_0 - H = B\left(\frac{1}{\mu_0} - \frac{1}{\mu}\right).$$

Let us again consider a small volume element and suppose that it has a magnetic dipole moment per unit volume, M. We saw in Section 6-8 that a current I enclosing an area S was equivalent to a magnetic dipole of moment IS. Applying this idea to Fig. 8-1, we find that a dipole moment per unit volume M throughout the current element would give a dipole moment of $M\,\Delta A\,\Delta l$ for the volume element, which is equivalent to a current $M\,\Delta l$ flowing around the surface. If the magnetization M is constant throughout the medium, and if we build up the medium with a number of volume elements, then the currents will cancel in the interior of the medium, and only a surface current of magnitude M per unit length will remain at the boundary. Identifying this current with the current change deduced above, we obtain

$$\frac{I_0 - I}{l} = M = B\left(\frac{1}{\mu_0} - \frac{1}{\mu}\right),$$

$$\therefore M = \frac{1}{\mu_0}B - H,$$

or

$$B = \mu_0(H + M).$$

For the simplest types of medium, B, H, and M are all in the same direction, and thus we have the vector equation

$$\mathbf{B} = \mu_0(\mathbf{H} + \mathbf{M}).$$

Note that this relation is slightly different from the corresponding one for the electric case. Dividing the scalar equation through by $\mu_0 H$ gives

$$\mu_r = 1 + \chi_m,$$

where $\mu_r = \mu/\mu_0$, the relative permeability, and $\chi_m = M/H$, the magnetic susceptibility.

Atomic theory again helps us visualize how these effects come about. A changing magnetic field in a conducting body will cause currents to flow, as we have already seen. Extending this idea to the model for the atom consisting of electrons traveling in orbits around the nucleus of the atom, we see that the establishing of a magnetic field will perturb these orbits, and since a moving charge is equivalent to a current, the perturbation of the motion is equivalent to additional currents flowing. Theories of magnetism have been developed in considerable detail, but we shall not pursue the matter further here. We should note, however, that unlike the dielectric constant, which for material media is always greater than unity, the relative permeability can be less than unity. Materials having a negative magnetic susceptibility are called *diamagnetic.* Those with relative permeability greater than one are known as *paramagnetic.* There is also a special class of paramagnetic materials (e.g. iron) which has very high values of relative permeability, and this class is known as *ferromagnetic.*

We shall now carry through the more general argument corresponding to that given for the electric case. From Section 6–8, for the magnitude of the vector potential for a current loop situated in vacuum, we have

$$\frac{\mu_0 I a^2 \sin\theta}{4r^2} = \frac{\mu_0 I(\pi a^2)\sin\theta}{4\pi r^2} = \frac{\mu_0 IS \sin\theta}{4\pi r^2}.$$

where IS is the magnetic moment. If this is caused by a magnetization M, then the vector potential due to volume $d\tau$ is

$$dA = \frac{\mu_0 M \sin\theta}{4\pi r^2}\, d\tau.$$

From Section 6–8 we know that this may be written in vector form as

$$d\mathbf{A} = \frac{\mu_0}{4\pi}\frac{\mathbf{M}\times\mathbf{u}_r}{r^2}\, d\tau,$$

and if we transfer the origin to the point at which we calculate the potential, as we did before, we have

$$d\mathbf{A} = -\frac{\mu_0}{4\pi}\frac{\mathbf{M}\times\mathbf{u}_r'}{r^2}\, d\tau = \frac{\mu_0}{4\pi}\mathbf{M}\times \operatorname{grad}\frac{1}{r}\, d\tau.$$

From the vector identity,

$$\text{curl}\, \phi\mathbf{A} = \phi\, \text{curl}\, \mathbf{A} - \mathbf{A} \times \text{grad}\, \phi,$$

we have

$$d\mathbf{A} = \frac{\mu_0}{4\pi}\left(\frac{1}{r}\, \text{curl}\, \mathbf{M} - \text{curl}\, \frac{\mathbf{M}}{r}\right),$$

and

$$\mathbf{A} = \frac{\mu_0}{4\pi} \iiint \frac{1}{r}\, \text{curl}\, \mathbf{M}\, d\tau - \frac{\mu_0}{4\pi} \iiint \text{curl}\, \frac{\mathbf{M}}{r}\, d\tau.$$

From Chapter 6 we see that the first term represents the vector potential due to a current distribution $\mathbf{J}' = \text{curl}\, \mathbf{M}$. By analogy with the electric case we would expect the second term to give the surface currents, and hence we seek a way of transforming this integral into a surface integral. In the vector identity

$$\text{div}\, (\mathbf{A} \times \mathbf{B}) = \mathbf{B} \cdot \text{curl}\, \mathbf{A} - \mathbf{A} \cdot \text{curl}\, \mathbf{B},$$

we let $\mathbf{A}$ be a constant vector so that curl $\mathbf{A} = 0$. Thus

$$\text{div}\, (\mathbf{A} \times \mathbf{B}) = -\mathbf{A} \cdot \text{curl}\, \mathbf{B}.$$

Using the divergence theorem, we obtain

$$\iiint \text{div}\, (\mathbf{A} \times \mathbf{B})\, d\tau = \iint \mathbf{A} \times \mathbf{B} \cdot d\mathbf{S} = \iint \mathbf{A} \cdot \mathbf{B} \times d\mathbf{S},$$

by using the rules for vector triple products. Since $\mathbf{A}$ is a constant, it may be brought outside the integral sign, and so

$$\iiint \text{div}\, (\mathbf{A} \times \mathbf{B})\, d\tau = \mathbf{A} \cdot \iint \mathbf{B} \times d\mathbf{S}.$$

But we also had

$$\text{div}\, (\mathbf{A} \times \mathbf{B}) = -\mathbf{A} \cdot \text{curl}\, \mathbf{B},$$

and thus

$$\mathbf{A} \cdot \iiint \text{curl}\, \mathbf{B}\, d\tau = -\mathbf{A} \cdot \iint \mathbf{B} \times d\mathbf{S} = \mathbf{A} \cdot \iint d\mathbf{S} \times \mathbf{B}.$$

Now, $\mathbf{A}$ can be any constant vector. If we let $\mathbf{A} = \mathbf{u}_x$, then this formula shows that the x-components of the two vectors $\iiint$ curl $\mathbf{B}\, d\tau$ and $\iint d\mathbf{S} \times \mathbf{B}$ are equal, since the dot product of a vector and a unit vector gives the magnitude of the component in that direction. Similarly, if we let $\mathbf{A} = \mathbf{u}_y$ we would show that the y-components are equal, and if $\mathbf{A} = \mathbf{u}_z$ that the z-components are equal. If all three components of a vector are equal, the vectors themselves are equal, and hence

$$\iiint \operatorname{curl} \mathbf{B}\, d\tau = \iint d\mathbf{S} \times \mathbf{B} = - \iint \mathbf{B} \times d\mathbf{S}.$$

Using this to transform the integral in the relation above, we obtain

$$\mathbf{A} = \frac{\mu_0}{4\pi} \iiint \frac{1}{r} \operatorname{curl} \mathbf{M}\, d\tau + \frac{\mu_0}{4\pi} \iint \frac{\mathbf{M} \times d\mathbf{S}}{r}.$$

Let us consider the first term in more detail. This, as we have said, shows that the dipole-moment density distribution is equivalent to a current density $\mathbf{J}' = \operatorname{curl} \mathbf{M}$. In vacuum, $\mathbf{B} = \mu_0 \mathbf{H}_0$, and if there is a current distribution $\mathbf{J}$, then

$$\operatorname{curl} \mathbf{H}_0 = \operatorname{curl} \frac{\mathbf{B}}{\mu_0} = \mathbf{J}.$$

In the presence of our dipole-moment distribution, this becomes

$$\operatorname{curl} \frac{\mathbf{B}}{\mu_0} = \mathbf{J} + \mathbf{J}' = \mathbf{J} + \operatorname{curl} \mathbf{M},$$

or

$$\operatorname{curl} \left(\frac{\mathbf{B}}{\mu_0} - \mathbf{M} \right) = \mathbf{J}.$$

In the presence of a medium with permeability μ, we have

$$\operatorname{curl} \mathbf{H} = \operatorname{curl} \frac{\mathbf{B}}{\mu} = \mathbf{J}.$$

Thus the effect of a medium with permeability μ is the same as that of a dipole-moment density distribution $\mathbf{M}$ in vacuum given by

$$\frac{\mathbf{B}}{\mu_0} - \mathbf{M} = \frac{\mathbf{B}}{\mu} = \mathbf{H},$$

$$\therefore \mathbf{B} = \mu_0(\mathbf{H} + \mathbf{M}).$$

The second term again gives the surface effect. This shows that there is a surface current J_s of magnitude M flowing at the surface. Let us consider the direction in the surface perpendicular to $\mathbf{M}$. If the magnetic field in this direction is $H_\perp$ (see Section 4–6) and there is a component of surface current J_s in this direction, then the boundary condition is

$$H_\perp^{(1)} - H_\perp^{(2)} = \frac{1}{\mu_0} B_\perp^{(1)} - \frac{1}{\mu_0} B_\perp^{(2)} = J_s.$$

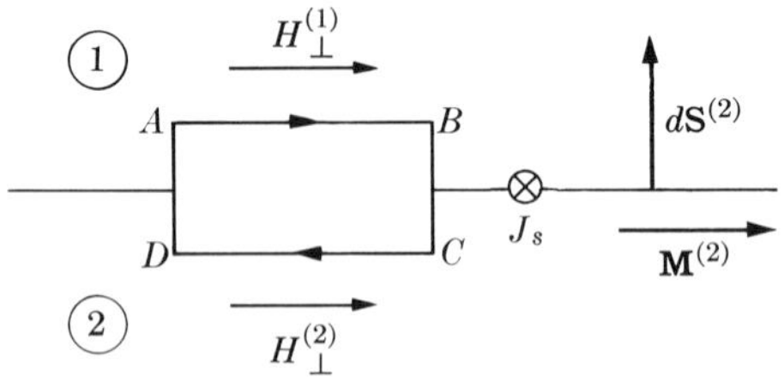

FIG. 8–4. The effect of the surface-current distribution.

In the presence of the dipole-moment distribution, this expression becomes

$$\frac{1}{\mu_0} B_\perp^{(1)} - \frac{1}{\mu_0} B_\perp^{(2)} = J_s - J_s',$$

the sign being obtained by reference to Fig. 8–4. This follows Fig. 4–2, but we now have an additional term. The positive direction for $H_\perp^{(2)}$ is to the right, and hence positive $\mathbf{M}^{(2)}$ is also to the right, as shown. The positive direction of $d\mathbf{S}$ is outward, and hence $\mathbf{M}^{(2)} \times d\mathbf{S}^{(2)}$ is in the opposite direction to $\mathbf{J}_s$. Hence

$$\frac{1}{\mu_0} B_\perp^{(1)} - \frac{1}{\mu_0} B_\perp^{(2)} = J_s - M,$$

or

$$\frac{1}{\mu_0} B_\perp^{(1)} - \left[\frac{1}{\mu_0} B_\perp^{(2)} - M\right] = J_s.$$

Extending this result to the case where medium 1 has permittivity μ_1 and medium 2 permittivity μ_2, we would have a magnetization vector in each medium, and the relation would be

$$\left[\frac{1}{\mu_0} B_\perp^{(1)} - M^{(1)}\right] - \left[\frac{1}{\mu_0} B_\perp^{(2)} - M^{(2)}\right] = J_s.$$

But from Section 4–6 we see that the boundary condition is

$$H_\perp^{(1)} - H_\perp^{(2)} = J_s.$$

Hence $(1/\mu_0)B_\perp - M$ obeys the same law as $H_\perp$. Here again the derivation is rather involved. It shows that the effect of a medium with permeability μ on the field is the same as that of a current distribution situated in vacuum composed of a volume distribution $\mathbf{J}' = \text{curl}\,\mathbf{M}$ and a surface distribution $\mathbf{J}_s' = \mathbf{M} \times \mathbf{n}$, where $\mathbf{n}$ is the unit outward-drawn normal vector, and $\mathbf{M}$ is given by $\mu\mathbf{H} = \mu_0(\mathbf{H} + \mathbf{M})$.

Problems

8–1 Consider a large block of dielectric with susceptibility χ. Suppose that the electric field in the block is **E**. Calculate the field inside a cavity in the material consisting of a long cylinder with axis parallel to **E** and with a small cross section ΔA.

8–2 Repeat Problem 8–1 for a disk-shaped cavity, given that the axis has length Δl and the cross section has area A.

8–3 Calculate the electron density needed to produce a value of zero for the refractive index of an ionized gas at 3 Mc/sec and at 3000 Mc/sec.

8–4 Repeat the calculation for the dielectric constant of an ionized gas including the motion of the heavy ions. Obtain the ratio of the contributions to the dielectric constant of the electrons and the ions.

8–5 An iron tube has length l, inner radius a, and outer radius b. The magnetization vector **M** in the iron is uniform and parallel to the axis of the tube. Calculate the field at the center of the tube, assuming l to be large in comparison with both a and b, by considering the equivalent pole strength. What is the field if l becomes infinite?

8–6 Repeat Problem 8–5, using the equivalent surface currents.

8–7 Repeat Problem 8–1 for a spherical cavity of radius a.

CHAPTER 9

Waves and Radiation

9–1 Introduction

In Chapter 5 we saw that Maxwell's equations and the constitutive relations provide five equations among five unknowns, which, along with the boundary conditions for a particular case, are sufficient to obtain a solution. In this chapter we shall show how a solution of these equations leads to waves, and how these can be excited.

9–2 The use of complex fields

The five equations we must solve are the following:

$$-\operatorname{curl} \mathbf{E} = \frac{\partial \mathbf{B}}{\partial t}, \qquad \mathbf{J} = \sigma\mathbf{E},$$

$$\operatorname{curl} \mathbf{H} = \frac{\partial \mathbf{D}}{\partial t} + \mathbf{J}, \qquad \mathbf{D} = \epsilon\mathbf{E},$$

$$\mathbf{B} = \mu\mathbf{H}.$$

We first make some simplifying assumptions concerning the medium. We suppose that σ, ϵ, and μ are scalar constants, so that $\mathbf{J}$ is in the same direction as $\mathbf{E}$, $\mathbf{D}$ is in the same direction as $\mathbf{E}$, and $\mathbf{B}$ is in the same direction as $\mathbf{H}$. The medium is therefore of the type we have termed isotropic. If σ is a constant (not dependent on $\mathbf{E}$), then if $\mathbf{E}$ is doubled, so is $\mathbf{J}$, and thus the medium is linear. (Similar results are obtained for the other two constitutive relations.) The constant nature of σ also means that it is not a function of x, y, and z (or whatever position coordinates we choose to employ). Hence the conductivity is the same at all points in the medium. Such a medium is called *homogeneous* or, more precisely, is said to have a *homogeneous conductivity*. In the case considered here, we also have homogeneous permittivity and permeability. If we eliminate $\mathbf{J}$, $\mathbf{D}$, and $\mathbf{B}$ from Maxwell's equations by using the constitutive relations, we obtain

$$-\operatorname{curl} \mathbf{E} = \frac{\partial}{\partial t}(\mu\mathbf{H}), \qquad \operatorname{curl} \mathbf{H} = \frac{\partial}{\partial t}(\epsilon\mathbf{E}) + \sigma\mathbf{E},$$

and since μ and ϵ are constants (independent of time), we have

$$-\operatorname{curl} \mathbf{E} = \mu\frac{\partial \mathbf{H}}{\partial t}, \qquad \operatorname{curl} \mathbf{H} = \epsilon\frac{\partial \mathbf{E}}{\partial t} + \sigma\mathbf{E}.$$

We are left with two equations in the two unknowns **E** and **H**. However, these are partial differential equations involving the four partial differential operators $\partial/\partial x$, $\partial/\partial y$, $\partial/\partial z$ (from the curl operator), and $\partial/\partial t$. In circuit theory the fundamental equations also contain the $\partial/\partial t$ operator, but we remove this by assuming that all quantities vary as $e^{j\omega t}$, which leads to a simplification of the problem. In the same way we can introduce sinusoidal time-varying quantities into the field equations. If we suppose that each of the components of the electric field **E** varies sinusoidally with time, then we can write

$$E_x = (E_x)_p \cos(\omega t + \alpha_x),$$
$$E_y = (E_y)_p \cos(\omega t + \alpha_y),$$
$$E_z = (E_z)_p \cos(\omega t + \alpha_z),$$

where $(E_x)_p$, $(E_y)_p$, and $(E_z)_p$ are the peak values of the field components, which are, in general, functions of position (x, y, z) but not functions of time. Writing these in complex form, we have

$$E_x = \mathrm{Re}\,[(E_x)_p e^{j\alpha_x} e^{j\omega t}] = \mathrm{Re}\,\mathcal{E}_x e^{j\omega t}$$

where

$$\mathcal{E}_x = (E_x)_p e^{j\alpha_x},$$

which is a complex number. Corresponding relations may be obtained for $\mathcal{E}_y$ and $\mathcal{E}_z$. Hence

$$\mathbf{E} = \mathrm{Re}\,(\mathcal{E}_x e^{j\omega t})\mathbf{u}_x + \mathrm{Re}\,(\mathcal{E}_y e^{j\omega t})\mathbf{u}_y + \mathrm{Re}\,(\mathcal{E}_z e^{j\omega t})\mathbf{u}_z = \mathrm{Re}\,(\boldsymbol{\mathcal{E}} e^{j\omega t}),$$

where

$$\boldsymbol{\mathcal{E}} = \mathcal{E}_x \mathbf{u}_x + \mathcal{E}_y \mathbf{u}_y + \mathcal{E}_z \mathbf{u}_z.$$

Thus $\boldsymbol{\mathcal{E}}$ is a vector, and the magnitudes of its components $\mathcal{E}_x$, $\mathcal{E}_y$, and $\mathcal{E}_z$ are complex numbers. When we write these numbers explicitly in terms of their real and imaginary parts, we use the notation

$$\mathcal{E}_x = \mathcal{E}_x' + j\mathcal{E}_x''.$$

(Note that we reserve the symbol j for $\sqrt{-1}$. Many authors use **i**, **j**, **k** as unit vectors in Cartesian coordinates; here $\mathbf{u}_x$, $\mathbf{u}_y$, and $\mathbf{u}_z$ are used. Also, x and y are reserved for space coordinates—two of the three Cartesian coordinates x, y, z. Some authors use x- and y-suffixes to denote real and imaginary parts of complex numbers. Here prime and double prime are used.)

The new quantity $\boldsymbol{\mathcal{E}}$ which we have introduced is a complex field. It is a function of position but not time:

$$\mathbf{E}(x, y, z, t) = \mathrm{Re}\,[\boldsymbol{\mathcal{E}}(x, y, z)e^{j\omega t}].$$

Similarly, we introduce a complex $\mathcal{H}$. Substituting in the first curl equation, we obtain

$$-\text{curl}\,[\text{Re}\,(\mathcal{E}e^{j\omega t})] = \mu\frac{\partial}{\partial t}\text{Re}\,(\mathcal{H}e^{j\omega t}),$$

and we have

$$\frac{\partial}{\partial t}\text{Re}\,(\mathcal{H}e^{j\omega t}) = \frac{\partial}{\partial t}\text{Re}\,[(\mathcal{H}_x\mathbf{u}_x + \mathcal{H}_y\mathbf{u}_y + \mathcal{H}_z\mathbf{u}_z)e^{j\omega t}]$$

$$= \frac{\partial}{\partial t}[\text{Re}\,(\mathcal{H}_x e^{j\omega t})\mathbf{u}_x + \text{Re}\,(\mathcal{H}_y e^{j\omega t})\mathbf{u}_y + \text{Re}\,(\mathcal{H}_z e^{j\omega t})\mathbf{u}_z],$$

since the real part of the sum of complex numbers is equal to the sum of their real parts. Also, the differential of the real part is equal to the real part of the differential, and therefore

$$\frac{\partial}{\partial t}\text{Re}\,(\mathcal{H}_x e^{j\omega t})\mathbf{u}_x = \text{Re}\left[\frac{\partial}{\partial t}(\mathcal{H}_x e^{j\omega t})\right]\mathbf{u}_x = \text{Re}\,(j\omega\mathcal{H}_x e^{j\omega t})\mathbf{u}_x,$$

since $\mathcal{H}_x$ is not a function of time. Similar relations are obtained for the y- and z-components. Hence we have

$$\frac{\partial}{\partial t}\text{Re}\,(\mathcal{H}e^{j\omega t}) = \text{Re}\,(j\omega\mu\mathcal{H}e^{j\omega t}).$$

By writing curl in Cartesian components and using the same properties of complex numbers, we may show that

$$\text{curl}\,[\text{Re}(\mathcal{E}e^{j\omega t})] = \text{Re}\,[(\text{curl}\,\mathcal{E})e^{j\omega t}],$$

and therefore

$$\text{Re}\,[-(\text{curl}\,\mathcal{E})e^{j\omega t}] = \text{Re}\,(j\omega\mu\mathcal{H}e^{j\omega t}).$$

This equation is true for all values of t. Putting $t = 0$, we have

$$\text{Re}\,(-\text{curl}\,\mathcal{E}) = \text{Re}\,(j\omega\mu\mathcal{H}),$$

and putting $\omega t = \pi/2$, we obtain

$$\text{Im}\,(-\text{curl}\,\mathcal{E}) = \text{Im}\,(j\omega\mu\mathcal{H}).$$

But if the real and the imaginary parts of a complex number are equal, then the numbers themselves are equal, and thus

$$-\text{curl}\,\mathcal{E} = j\omega\mu\mathcal{H},$$

and similarly,

$$\text{curl}\,\mathcal{H} = (\sigma + j\omega\epsilon)\mathcal{E}.$$

Hence, by introducing sinusoidal time variation we introduce complex fields and remove one of the four differential operators from our equations.

9–3 Waves in a dielectric

If we suppose that our medium is a dielectric, then its conductivity is zero, and the equations of Section 9–2 become

$$-\text{curl}\, \boldsymbol{\mathcal{E}} = j\omega\mu\boldsymbol{\mathcal{H}}, \qquad \text{curl}\, \boldsymbol{\mathcal{H}} = j\omega\epsilon\boldsymbol{\mathcal{E}}.$$

From the first equation

$$-\text{curl curl}\, \boldsymbol{\mathcal{E}} = \text{curl}\, j\omega\mu\boldsymbol{\mathcal{H}} = j\omega\mu\, \text{curl}\, \boldsymbol{\mathcal{H}},$$

since the curl operator involves the operators $\partial/\partial x$, $\partial/\partial y$, and $\partial/\partial z$, and $j\omega\mu$ is not a function of x, y, and z. Hence we eliminate $\boldsymbol{\mathcal{H}}$ from the equations, and we have

$$-\text{curl curl}\, \boldsymbol{\mathcal{E}} = j\omega\mu \cdot j\omega\epsilon\boldsymbol{\mathcal{E}} = -\omega^2\mu\epsilon\boldsymbol{\mathcal{E}}.$$

The operation curl curl is one that we studied in Section 6–6, and we have

$$\text{curl curl}\, \boldsymbol{\mathcal{E}} = \text{grad div}\, \boldsymbol{\mathcal{E}} - \nabla^2\boldsymbol{\mathcal{E}}.$$

If we make the further assumption that there are no sources (no charges or currents) in the medium, then div $\mathbf{D} = 0$, from which, in conjunction with the above assumptions, it follows that div $\boldsymbol{\mathcal{E}} = 0$, and therefore curl curl $\boldsymbol{\mathcal{E}} = -\nabla^2\boldsymbol{\mathcal{E}}$ and

$$\nabla^2\boldsymbol{\mathcal{E}} + \omega^2\mu\epsilon\boldsymbol{\mathcal{E}} = 0.$$

The wave number k is defined by the relation

$$k^2 = \omega^2\mu\epsilon.$$

Hence the above equation for $\boldsymbol{\mathcal{E}}$ may be written

$$(\nabla^2 + k^2)\boldsymbol{\mathcal{E}} = 0.$$

From Section 6–6 we know that this equation means

$$(\nabla^2 + k^2)\mathcal{E}_x = 0, \qquad (\nabla^2 + k^2)\mathcal{E}_y = 0, \qquad (\nabla^2 + k^2)\mathcal{E}_z = 0.$$

A further simplification may be obtained by assuming that $\boldsymbol{\mathcal{E}}$ has only an x-component, that is, $\mathcal{E}_y = \mathcal{E}_z = 0$. In addition, we suppose that $\mathcal{E}_x$ is a function only of z; i.e., the x-component of $\boldsymbol{\mathcal{E}}$ has the same magnitude for all values of x and y and depends only on the value of z. Writing ∇^2 in Cartesian coordinates, we have

$$\frac{\partial^2\mathcal{E}_x}{\partial x^2} + \frac{\partial^2\mathcal{E}_x}{\partial y^2} + \frac{\partial^2\mathcal{E}_x}{\partial z^2} + k^2\mathcal{E}_x = 0.$$

But we have said that $\mathcal{E}_x$ does not depend on x and y, and therefore the first two terms are zero, and we have

$$\frac{d^2\mathcal{E}_x}{dz^2} + k^2\mathcal{E}_x = 0.$$

This is the equation of simple harmonic motion which has the solution

$$\mathcal{E}_x = ae^{jkz} + be^{-jkz} \quad \text{and} \quad \boldsymbol{\mathcal{E}} = (ae^{jkz} + be^{-jkz})\mathbf{u}_x.$$

It should be noted that the solution to this equation must satisfy the condition div $\boldsymbol{\mathcal{E}} = 0$ if it is to be a possible solution to our original problem. It may be readily verified that the above solution satisfies this condition.

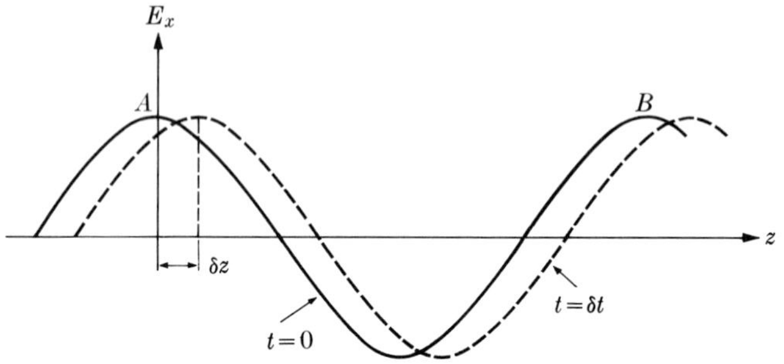

FIG. 9–1. A traveling wave.

Let us consider the case where $a = 0$ and b is real. Then

$$\mathbf{E} = \mathbf{u}_x E_x = \mathbf{u}_x \,\mathrm{Re}\,(be^{-jkz}e^{j\omega t}),$$

$$\therefore \mathbf{E} = \mathbf{u}_x b \cos(\omega t - kz).$$

At time $t = 0$, $\mathbf{E}_x$ varies sinusoidally with z and has a maximum at the origin (Fig. 9–1). At time δt, the maximum at the origin has moved to the new point δz given by $\omega\,\delta t - k\,\delta z = 0$ or $\delta z = (\omega/k)\,\delta t$. Similarly all other points on the curve have moved by the same amount, and the whole curve appears to have moved to the right, as shown in Fig. 9–1; i.e., the wave appears to be traveling and is consequently known as a *traveling wave*. It moves a distance δz in time δt and thus is moving with a velocity

$$\frac{\delta z}{\delta t} = \frac{\omega}{k},$$

which is known as the *phase velocity*. Substituting for k we have

$$v_\phi = \frac{\omega}{\sqrt{\omega^2\mu\epsilon}} = \frac{1}{\sqrt{\mu\epsilon}}.$$

Thus v_ϕ is independent of frequency. A medium in which the phase velocity is not a function of frequency is said to be *nondispersive*.

For a vacuum the permeability is $4\pi \times 10^{-7}$ h/m by definition, and the permittivity is approximately $(1/36\,\pi) \times 10^{-9}$ f/m. Substituting, we find that the phase velocity in vacuum is

$$c = \frac{1}{\sqrt{\mu_0 \epsilon_0}} = 3 \times 10^8 \text{ m/sec},$$

or more accurately, 2.998×10^8 m/sec. This is equal to the velocity of light, a fact first noticed by Maxwell.

We can attach a physical significance to k in this particular case. For a given value of t (say $t = 0$), E_x varies like $\cos kz$. If kz increases by 2π, then the cosine passes through a complete cycle such as that between A and B in Fig. 9–1. The distance AB is the wavelength in the medium λ. Hence $k\lambda = 2\pi$, or $k = 2\pi/\lambda$.

In the expression $\cos(\omega t + \alpha)$ we refer to α as the *phase angle*, and if two sinusoidally varying quantities have the same value for α, they are said to be *in phase* or to have *equal phase angles*. This implies that both quantities pass through their maximum, minimum, and zero values at the same instant. Since $\mathbf{E}$ is a vector field, then at each point in space we have a vector which, in this case, has only an x-component E_x which is a sinusoidal function of time. The phase angle of E_x is given by $(-kz)$, and for a given value of z, $(-kz) = \text{const}$, and all the oscillations are in phase. But $z = \text{const}$ defines a plane, and thus at all points on any plane parallel to the xy-plane (namely, $z = \text{const}$) the oscillations are in phase. Because the surface over which all phase angles are equal is a plane, the wave is referred to as a *plane wave*. We could plot a diagram of the type shown in Fig. 9–1 for each value of x and y. For a given value of t, each of these curves would have a maximum at the same value of z. As time increases all these curves move together, and hence the crests of all these curves move at the same velocity. In other words, the planes containing the crests of the curves (or any other particular point on the curve such as the troughs or the zero crossings) propagate at the phase velocity.

The amplitude of the oscillation is b at all points. If the amplitude of the wave is the same at all points where the phase angles are equal, the wave is said to be a *uniform wave*. In this case, the surfaces of equal phase angle are planes parallel to the xy-plane, and therefore a sufficient condition to ensure a uniform wave is that the amplitude is not dependent on x or y.

Another property of the wave is its polarization. Let us again consider the relation

$$\mathbf{E} = \mathbf{u}_x b \cos(\omega t - kz),$$

and choose some particular point in space, i.e., specify x, y, and z. Then in the above relation z is a constant. We investigate the behavior of $\mathbf{E}$ as time varies. We see that it is always pointing in the x-direction, and its magnitude varies sinusoidally. The tip of the vector representing $\mathbf{E}$ at this point will therefore trace a line. By definition, the curve traced by the tip of the $\mathbf{E}$-vector as time varies describes the polarization of the wave. Thus the wave is said to be linearly polarized, with the $\mathbf{E}$-vector in the x-direction.

To illustrate the idea of polarization further we remove the restriction that $E_y = 0$. By solving the equation for E_y in the same way as we did for E_x we get

$$\mathcal{E}_y = ce^{jkz} + de^{-jkz}.$$

We assumed above, by implication, that a and b were real. We recognize that, in general, they can be complex, and we must allow for this possibility. We again take $c = 0$ and let d be a complex number, and thus

$$E_y = \text{Re}\,(|d|e^{j\delta}e^{-jkz}e^{j\omega t}) = |d| \cos(\omega t - kz + \delta).$$

By using the principle of superposition (we assume the medium to be linear), we obtain

$$\mathbf{E} = \mathbf{u}_x b \cos(\omega t - kz) + \mathbf{u}_y |d| \cos(\omega t - kz + \delta).$$

Let us consider the case where $b = |d| = 1$, and $\delta = -(\pi/2)$. Then we have

$$\mathbf{E} = \mathbf{u}_x \cos(\omega t - kz) + \mathbf{u}_y \sin(\omega t - kz),$$

from which we see that the magnitude of $\mathbf{E}$ is unity, and the angle between the x-axis and $\mathbf{E}$ is

$$\tan^{-1}\frac{E_y}{E_x} = \tan^{-1}\left[\frac{\sin(\omega t - kz)}{\cos(\omega t - kz)}\right] = \omega t - kz.$$

Thus, for a fixed point in space (z = const), this angle increases linearly with time. The tip of the $\mathbf{E}$-vector therefore traces a circle, and the wave is circularly polarized.

Thus far we have considered only the $\mathbf{E}$-vector. By eliminating $\boldsymbol{\mathcal{E}}$ from the two equations at the end of the previous section, we obtain an equation for $\boldsymbol{\mathcal{H}}$ which is identical to that obeyed by $\boldsymbol{\mathcal{E}}$. Thus $\mathbf{H}$ has the same features that we discussed for $\mathbf{E}$. We must now determine the relation between $\mathbf{E}$ and $\mathbf{H}$. If we choose our solution by specifying special conditions for $\mathbf{E}$, then we must use the resulting solution for $\mathbf{E}$ to obtain $\mathbf{H}$, which is simply done from the equation

$$-\text{curl}\,\boldsymbol{\mathcal{E}} = j\omega\mu\boldsymbol{\mathcal{H}}.$$

For example, let us suppose that we take the solution for $\boldsymbol{\mathcal{E}}$ to be

$$\boldsymbol{\mathcal{E}} = \mathbf{u}_x b e^{jkz}.$$

Then

$$jkbe^{-jkz}\mathbf{u}_y = j\omega\mu\boldsymbol{\mathcal{H}},$$

$$\therefore \boldsymbol{\mathcal{H}} = (k/\omega\mu)be^{-jkz}\mathbf{u}_y = b\sqrt{\epsilon/\mu}\, e^{-jkz}\mathbf{u}_y,$$

and

$$\mathbf{H} = b\sqrt{\epsilon/\mu}\cos(\omega t - kz)\mathbf{u}_y.$$

Thus **H** has only a y-component, and it oscillates in phase with **E** at all points (i.e., they both pass through their maximum value at the same instant of time). Expressed independently of the coordinate system, we can say that this wave has linearly polarized **E**- and **H**-vectors directed so that **E**, **H**, and the direction of propagation form a right-handed set of mutually perpendicular directions and $E/H = \sqrt{\mu/\epsilon}$.

We have seen that the dimensions of μ are inductance per unit length and those of ϵ are capacitance per unit length. Hence $\sqrt{\mu/\epsilon}$ has the dimensions of impedance. Note that this is a property of the medium, which is referred to as the intrinsic impedance of the medium, η:

$$\eta = \sqrt{\mu/\epsilon}.$$

For a vacuum it has the value 377 ohms. The intrinsic impedance of a medium plays a role in field theory similar to that of the characteristic impedance in transmission line theory. The term "intrinsic" is reserved for the parameters of the uniform plane traveling wave.

9–4 Standing waves

Let us consider the general solution arrived at for $\mathcal{E}_x$ in the last section, namely,

$$\mathcal{E}_x = ae^{jkz} + be^{-jkz},$$

and obtain another special case by taking a real and $a = b$. Thus

$$\mathcal{E}_x = a(e^{jkz} + e^{-jkz}) = 2a \cos kz.$$

Hence

$$\mathbf{E} = \text{Re}\,(2a \cos kz\, e^{j\omega t})\mathbf{u}_x, \qquad \text{and} \qquad \mathbf{E} = 2a \cos kz \cos \omega t\, \mathbf{u}_x.$$

The amplitude factor $2a$ is, of course, arbitrary. This expression gives a type of field which is quite different from that obtained in the last section. We note that the phase angle is everywhere zero; i.e., the oscillations are in phase throughout all space. But the amplitude of the oscillation is now a function of z. At $t = 0$ the variation of E_x with z is shown in Fig. 9–2. At $t = \delta t$ the varia-

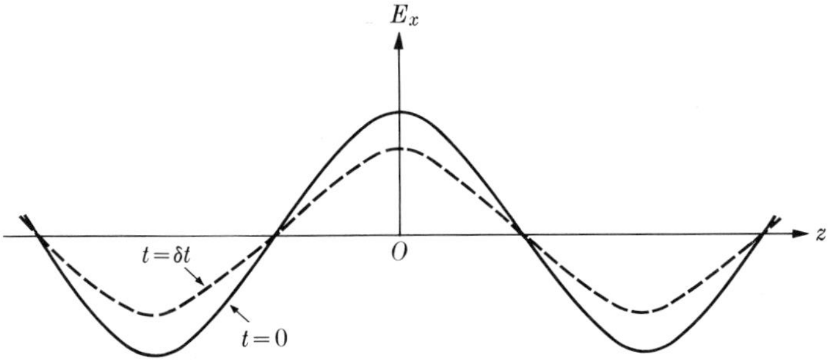

FIG. 9–2. A standing wave.

tion is still like cos kz (with a maximum at $z = 0$), but the amplitude of the cosinusoid (dashed curve) is reduced by a factor cos $(\omega\,\delta t)$. The wave does not propagate, and is known as a *standing wave*. Note that we cannot define a surface over which the phase angles are equal since they are equal everywhere. If we take the general case $a \neq b$, then the resultant wave has both a standing and a propagating component. Here the equal phase-angle surfaces are planes just as they are in the case of the pure traveling wave. It is convenient to consider the pure standing wave as a limiting case and view it as a plane wave whose amplitude is $2a \cos kz$. The equal phase-angle surfaces are defined by $z =$ const, and hence the amplitude over these surfaces is constant, and the wave is therefore uniform. The wave is also linearly polarized with the **E**-vector in the x-direction as before.

We obtain **H** via the equations

$$-\text{curl}\,\boldsymbol{\mathcal{E}} = j\omega\mu\boldsymbol{\mathcal{H}},$$

$$2ak \sin kz\,\mathbf{u}_y = j\omega\mu\boldsymbol{\mathcal{H}},$$

$$\boldsymbol{\mathcal{H}} = \frac{2a}{j}\frac{1}{\eta}\sin kz\,\mathbf{u}_y = 2a\frac{1}{\eta}\sin kz\,e^{-j(\pi/2)}\mathbf{u}_y,$$

$$\mathbf{H} = \text{Re}\,[2a\frac{1}{\eta}\sin kz\,e^{-j(\pi/2)}e^{j\omega t}\mathbf{u}_y]$$

$$= 2a\frac{1}{\eta}\sin kz\cos\left(\omega t - \frac{\pi}{2}\right)\mathbf{u}_y.$$

When $t = 0$, $\mathbf{H} = 0$, and thus at the instant of time when **E** is a maximum, **H** is zero and vice versa; **E** and **H** are in phase quadrature. In addition, **H** varies like sin kz, whereas **E** varies like cos kz, and hence at points where the amplitude of the **E**-oscillations are a maximum the amplitude of the **H**-oscillations are zero. The vector directions of **E** and **H** are the same as in the case of the propagating wave, **E** being in the x-direction and **H** in the y-direction.

9–5 Power flow

The waves considered above contain an **E**-field and an **H**-field, and, as we have seen, energy is associated with these fields, the energy density in the electric field being $\frac{1}{2}\epsilon E^2$ and in the magnetic field $\frac{1}{2}\mu H^2$. If we consider a loss-free dielectric ($\sigma = 0$), then the fields are related by the equations

$$-\text{curl}\,\mathbf{E} = \mu\frac{\partial\mathbf{H}}{\partial t}, \qquad \text{curl}\,\mathbf{H} = \epsilon\frac{\partial\mathbf{E}}{\partial t}.$$

From these we obtain

$$\mathbf{H}\cdot\text{curl}\,\mathbf{E} - \mathbf{E}\cdot\text{curl}\,\mathbf{H} = -\mathbf{H}\cdot\mu\frac{\partial\mathbf{H}}{\partial t} - \mathbf{E}\cdot\epsilon\frac{\partial\mathbf{E}}{\partial t} = -\frac{\partial}{\partial t}(\tfrac{1}{2}\mu H^2 + \tfrac{1}{2}\epsilon E^2).$$

By a vector identity the left-hand side is div $(\mathbf{E} \times \mathbf{H})$, and therefore

$$\text{div}\,(\mathbf{E} \times \mathbf{H}) = -\frac{\partial}{\partial t}\left(\tfrac{1}{2}\mu H^2 + \tfrac{1}{2}\epsilon E^2\right).$$

Integrating throughout a volume and using the divergence theorem, we obtain

$$\iiint \text{div}\,(\mathbf{E} \times \mathbf{H})\,d\tau = \oiint \mathbf{E} \times \mathbf{H} \cdot d\mathbf{S} = -\frac{\partial}{\partial t}\iiint \left(\tfrac{1}{2}\mu H^2 + \tfrac{1}{2}\epsilon E^2\right) d\tau,$$

where the surface integral is taken over the surface bounding the volume through which the volume integral is taken. The integral on the right-hand side represents the energy density integrated throughout the volume and is thus the total energy stored in it. The right-hand side of the equation therefore represents the rate of decrease of energy within the volume. Since the medium is lossless, this energy cannot be dissipated, and by the law of conservation of energy, it must therefore leave the volume by flowing out across the bounding surface. The surface integral is taken over this bounding surface, and if we interpret $\mathbf{E} \times \mathbf{H}$ to be the power flow per unit area, then this integral would represent the total power flowing outward across the surface, which is equal to the rate of decrease of stored energy inside the surface. The vector $\mathbf{E} \times \mathbf{H}$ is known as the *Poynting vector* and is usually represented by $\mathbf{S}$.

When the medium is not lossless, we have

$$-\text{curl}\,\mathbf{E} = \mu\frac{\partial \mathbf{H}}{\partial t}, \qquad \text{curl}\,\mathbf{H} = \epsilon\frac{\partial \mathbf{E}}{\partial t} + \sigma\mathbf{E}.$$

Following the same steps as before and changing the sign, we obtain

$$-\oiint \mathbf{E} \times \mathbf{H} \cdot d\mathbf{S} = \frac{\partial}{\partial t}\iiint \left(\tfrac{1}{2}\mu H^2 + \tfrac{1}{2}\epsilon E^2\right) d\tau + \iiint \sigma E^2\,d\tau,$$

where the additional term $\iiint \sigma E^2 d\tau$, which appears on the right-hand side, represents the rate of power dissipation integrated throughout the volume. The equation may now be interpreted in terms of the law of conservation of energy. The left-hand side represents the rate of power flow into the volume, which is equal to the rate of increase of stored energy within the volume plus the rate of power dissipation within the volume.

Let us consider the linearly polarized traveling wave of the last section. The fields are given by

$$\mathbf{E} = b\cos(\omega t - kz)\,\mathbf{u}_x = E_0\cos(\omega t - kz)\,\mathbf{u}_x,$$

$$\mathbf{H} = b\frac{1}{\eta}\cos(\omega t - kz)\,\mathbf{u}_y = H_0\cos(\omega t - kz)\,\mathbf{u}_y.$$

The energy density in the electric field is given by

$$\tfrac{1}{2}\epsilon E^2 = \tfrac{1}{2}\epsilon E_0^2 \cos^2 (\omega t - kz),$$

and its time average is $\frac{1}{4}\epsilon E_0^2$. (Note that E_0 is the peak value of the field.) Similarly the time-average energy in the magnetic field is $\frac{1}{4}\mu H_0^2$. We also have

$$\tfrac{1}{4}\mu H_0^2 = \tfrac{1}{4}\mu \left(\frac{1}{\eta} E_0\right)^2 = \tfrac{1}{4}\epsilon E_0^2,$$

which states that the time-average energies in the electric and magnetic fields are equal.

The power flow density $\mathbf{S}$ is given by

$$\mathbf{S} = E_0 H_0 \cos^2 (\omega t - kz)\, \mathbf{u}_x \times \mathbf{u}_y,$$

and the time-average flow is

$$\overline{\mathbf{S}} = \frac{1}{2} E_0 H_0 \mathbf{u}_z = \frac{1}{2}\frac{1}{\eta} E_0^2 \mathbf{u}_z.$$

This expression shows that the power flow is in the z-direction, the same direction as that in which the wave travels.

Now let us consider the standing wave, where as we saw above

$$\tfrac{1}{2}\epsilon E^2 = \tfrac{1}{2}\epsilon E_0^2 \cos^2 kz \cos^2 \omega t,$$

and

$$\tfrac{1}{2}\mu H^2 = \tfrac{1}{2}\mu H_0^2 \sin^2 kz \sin^2 \omega t = \tfrac{1}{2}\epsilon E_0^2 \sin^2 kz \sin^2 \omega t.$$

The power flow density is given by

$$\mathbf{S} = E_0 \cos kz \cos \omega t \,\frac{1}{\eta} E_0 \sin kz \sin \omega t = \frac{1}{4}\frac{1}{\eta} E_0^2 \sin 2kz \sin 2\omega t.$$

Hence the power flow density varies like $\sin 2\omega t$. The time-average power flow is zero, and the power oscillates back and forth at twice the wave frequency. When the energy in the magnetic field is a maximum, that in the electric field is zero and vice versa. Also, the positions along the z-axis where the magnetic field stores the energy are different from those where the electric field stores it, one varying as $\cos^2 kz$ and the other as $\sin^2 kz$. We can now interpret what is happening. At time $t = 0$ all the energy is stored in the electric field, and its density varies as $\cos^2 kz$ in the z-direction. As time progresses, this energy is transferred to the magnetic field, and at $t = \pi/2\omega$ it has all been transferred. The energy density now varies as $\sin^2 kz$, and hence there must have been a power flow in the z-direction to achieve this redistribution, a fact borne out by the Poynting vector. As time continues, the energy is transferred back to the electric field, and the power flow reverses. The energy is continually transferred back and forth between the electric and magnetic fields, and the time-average power flow is zero. If we take the average of the energy densities over z, we have

$$w_e = \tfrac{1}{4}\epsilon E_0^2 \cos^2 \omega t, \qquad w_m = \tfrac{1}{4}\epsilon E_0^2 \sin^2 \omega t,$$

and

$$w_e + w_m = \tfrac{1}{4}\epsilon E_0^2,$$

The sum of the energies is therefore constant at all times, as it should be from the law of conservation of energy. We may compare this oscillation of the stored energy between the electric and magnetic fields with the oscillation of the energy between the capacitor and the inductor in a tuned circuit.

We have just seen that the traveling wave transports energy, but the standing wave does not. In general, we can have a superposition of standing and traveling waves, and the ratio of the maximum amplitude of the oscillation to the minimum is, by definition, the standing wave ratio (SWR). Thus, if we wish to obtain maximum transfer of power, we need to make the SWR as close to unity as possible. The most common engineering application of this idea occurs in connection with waves on lines where it is desired to obtain the maximum transfer of energy between a source and a load.

9–6 Absorption

In this section we remove the restriction of zero conductivity of the medium; our equations are then

$$-\text{curl}\ \mathcal{E} = j\omega\mu\mathcal{H}, \qquad \text{curl}\ \mathcal{H} = (\sigma + j\omega\epsilon)\mathcal{E}.$$

We proceed as before and obtain

$$\nabla^2\mathcal{E} - j\omega\mu(\sigma + j\omega\epsilon)\mathcal{E} = 0.$$

A more general definition is adopted for k, namely,

$$k^2 = -j\omega\mu(\sigma + j\omega\epsilon).$$

Note that this reduces to the previous definition for k in the special case $\sigma = 0$. We therefore have

$$(\nabla^2 + k^2)\mathcal{E} = 0,$$

and the solution can proceed as before, the only difference being that k^2 is now complex. The solution for the case of the conducting medium corresponding to the one obtained for dielectrics in the last section is

$$\mathcal{E} = (ae^{jkz} + be^{-jkz})\mathbf{u}_x.$$

Taking $a = 0$ and b real, and writing k in its real and imaginary parts, $k = k' - jk''$ (the minus sign being taken for convenience and being part of the definition of k''), we have

$$\boldsymbol{\mathcal{E}} = be^{-j(k'-jk'')z}\mathbf{u}_x = be^{-k''z}e^{-jk'z}\mathbf{u}_x,$$

from which we obtain

$$\mathbf{E} = \text{Re}\,(be^{-k''z}e^{-jk'z}e^{j\omega t})\mathbf{u}_x,$$

$$\therefore \mathbf{E} = be^{-k''z}\cos(\omega t - k'z)\cdot\mathbf{u}_x.$$

From Maxwell's equations we have

$$-\text{curl}\,\boldsymbol{\mathcal{E}} = j\omega\mu\boldsymbol{\mathcal{H}},$$

$$\therefore jkbe^{-jkz}\mathbf{u}_y = j\omega\mu\boldsymbol{\mathcal{H}},$$

or

$$\boldsymbol{\mathcal{H}} = \frac{k}{\omega\mu}be^{-jkz}\mathbf{u}_y.$$

But k is now complex, and we can write

$$k/\omega\mu = \sqrt{(\sigma + j\omega\epsilon)/j\omega\mu}.$$

This is the expression corresponding to that used to define the intrinsic impedance η. Here we have $\sigma + j\omega\epsilon$ where previously we had $j\omega\epsilon$. Note that for $\sigma = 0$ this reduces to $\sqrt{\epsilon/\mu}$. We can extend the idea of intrinsic impedance to include the case of complex impedance by the definition

$$\eta = \sqrt{j\omega\mu/(\sigma + j\omega\epsilon)}.$$

For present purposes it is convenient to write η in the form

$$\eta = |\eta|e^{j\delta}.$$

Then

$$\boldsymbol{\mathcal{H}} = \frac{b}{|\eta|}e^{-j\delta}e^{-jk'z}e^{-k''z}\mathbf{u}_y,$$

and

$$\mathbf{H} = \frac{b}{|\eta|}e^{-k''z}\cos(\omega t - k'z - \delta)\cdot\mathbf{u}_y.$$

From the expression for **E** we see that this solution represents a wave traveling in the z-direction with a phase velocity ω/k'. It is also a plane wave by the same arguments as used before. Its amplitude varies like $e^{-k''z}$, and thus the amplitude of the wave decreases in the direction of propagation. The equal phase-angle surfaces are the planes $z = \text{const}$, and we see that for constant z the amplitude is constant and thus the wave is a uniform wave. The polarization is the same as before. The **H**-vector is still perpendicular to the **E**-vector, **E** being in the x-direction and **H** in the y-direction. Thus the spatial relationship is unchanged; however, instead of the two oscillating in phase, the time relationship is modified by the introduction of the phase angle δ.

Let us now consider a highly conducting medium such as a metal, where $\sigma \gg j\omega\epsilon$. We then have

$$\eta = \sqrt{j\omega\mu/\sigma} = \sqrt{\omega\mu/\sigma}\, e^{j(\pi/4)} \qquad \text{and} \qquad k = \sqrt{-j\omega\mu\sigma} = \sqrt{\omega\mu\sigma/2}\,(1 - j).$$

Hence

$$\mathbf{E} = be^{-\sqrt{\omega\mu\sigma/2}\, z} \cos\,(\omega t - \sqrt{\omega\mu\sigma/2}\, z),$$

and

$$\mathbf{H} = b\sqrt{\sigma/\omega\mu}\, e^{-\sqrt{\omega\mu\sigma/2}\, z} \cos\,(\omega t - \sqrt{\omega\mu\sigma/2}\, z - \pi/4)\,\cdot$$

Here the **H**-vector lags the **E**-vector by $\pi/4$. For the phase velocity we have

$$v_\phi = \omega/k' = \sqrt{2\omega/\mu\sigma}.$$

This is a function of frequency, and thus the medium is dispersive. We also note that the phase velocity increases with frequency and decreases as the conductivity or the permeability increases. The wavelength is given by

$$\lambda_m = 2\pi\sqrt{2/\omega\mu\sigma}.$$

The penetration depth δ is defined as the distance which the wave must travel to be reduced in amplitude to $1/e$ times its initial value, that is, $k''\,\delta = 1$. Hence

$$\delta = 1/k'' = \sqrt{2/\omega\mu\sigma}.$$

Note that δ decreases as frequency, permeability, or conductivity increases. For copper, $\sigma \simeq 5 \times 10^7$ mho/m, and the relative permeability is approximately 1 so that $\mu = \mu_0 = 4\pi \times 10^{-7}$ h/m. Taking a frequency of 1 Mc/sec, we have

$$v_\phi = \sqrt{2\omega/\mu\sigma} \simeq 440 \text{ m/sec},$$

$$\lambda_m = v_\phi/f \simeq 4.4 \times 10^{-4} \text{ m} = 0.44 \text{ mm},$$

$$\delta = \sqrt{2/\omega\mu\sigma} \simeq 7.0 \times 10^{-5} \text{ m} = 0.07 \text{ mm}.$$

These parameters are drastically affected by the large value of the conductivity. The velocity in free space is 3×10^8 m/sec as compared to 4.4×10^2 m/sec in copper. The free-space wavelength is roughly 300 m compared to 0.44 mm in copper. The wave is very rapidly attenuated, and the fields are confined to the region near the surface of a conductor, a phenomenon known as *skin effect*.

We saw that in the case of dielectrics the energy was carried equally by the electric and magnetic fields. The ratio is given by

$$\frac{(1/2)\mu|H|^2}{(1/2)\epsilon|E|^2}.$$

For a nonmagnetic metal we may assume that $\mu \doteq \mu_0$. The permittivity of a conductor is a difficult quantity to measure, and it is customary to take a value of the same order as that for vacuum. Hence the ratio of magnetic to electric energy is given by

$$\frac{\mu_0}{\epsilon_0}\left(\frac{|H|}{|E|}\right)^2 = \eta_0^2\left(\frac{\sigma}{\omega\mu_0}\right).$$

Taking the same values as before for ω, μ_0, and σ, and setting $\eta_0 = 377$ ohms, this ratio becomes roughly 9×10^{11}. Thus almost all the energy resides in the magnetic field. The power flow density is then given by

$$\begin{aligned}\mathbf{S} &= \mathbf{E} \times \mathbf{H}\\ &= b^2\sqrt{\frac{\sigma}{\omega\mu}}\,e^{-2\sqrt{(1/2)\omega\mu\sigma}\,z}\cos\left(\omega t - \sqrt{\frac{\omega\mu\sigma}{2}}\,z\right)\cos\left(\omega t - \sqrt{\frac{\omega\mu\sigma}{2}}\,z - \frac{\pi}{4}\right)\mathbf{u}_z,\end{aligned}$$

and the time average, by

$$\overline{\mathbf{S}} = b^2\sqrt{\sigma/\omega\mu}\,e^{-2\sqrt{(1/2)\omega\mu\sigma}\,z}\,\tfrac{1}{2}\cos(\pi/4)\mathbf{u}_z.$$

Differentiating, we obtain

$$\frac{d\overline{\mathbf{S}}}{dz} = \tfrac{1}{2}b^2\sqrt{\sigma/2\omega\mu}\left(-2\sqrt{\tfrac{1}{2}\omega\mu\sigma}\right)e^{-2\sqrt{(1/2)\omega\mu\sigma}\,z}\,\mathbf{u}_z.$$

At $z = 0$ the power flow density is $\overline{\mathbf{S}}_0$, and at $z = \Delta z$ it is $\overline{\mathbf{S}}_0 + (d\overline{\mathbf{S}}/dz)_0\,\Delta z$. The difference is therefore $(d\overline{\mathbf{S}}/dz)_0\,\Delta z$ and is given by $-\frac{1}{2}b^2\sigma\,\Delta z\mathbf{u}_z$, and so there is less power flowing at $z = \Delta z$ than there is at $z = 0$, which we would expect since the amplitude is smaller. We know that the power dissipated per unit volume is given by σE^2 (Section 2–3), which in this case, at $z = 0$, is

$$\sigma b^2\cos^2(\omega t - k'z),$$

and the time-average power dissipated in a volume of unit cross-sectional area and thickness Δz is therefore $\frac{1}{2}b^2\sigma\,\Delta z$. Thus we see that the difference between the power crossing $z = 0$ and that crossing $z = \Delta z$ is exactly accounted for by the power dissipated in the medium between them. This power is absorbed by the medium and converted into heat. The coefficient k'' is accordingly referred to as the *absorption coefficient*. The coefficient k' determines how the phase angle of the oscillation changes with z, and is known as the *phase constant*.

9–7 Reflection of waves

When a wave is incident upon the interface between two media with different properties, part of the energy is reflected and part is transmitted. Let us consider, for example, a uniform plane wave incident upon the plane interface between media, as shown in Fig. 9–3. Note that the relative directions of the **E**- and **H**-vectors and the direction of propagation for each of the three waves has been drawn in accordance with the properties of the waves discussed in Section 9–3. The interface between the two media has been chosen to be the xy-plane, and we consider first the special case where the **E**-vector is parallel to the interface. The expression for a wave traveling in the z-direction and linerly polarized with **E** in the x-direction is obtained from Section 9–3 as

$$E_x = b \cos (\omega t - kz).$$

Here we need an expression for a linearly polarized wave traveling in a direction making an angle θ with the z-axis. Hence

$$E_x = b \cos (\omega t - kl)$$

and from Fig. 9–4 we see that

$$l = y \sin \theta + z \cos \theta,$$

so that

$$E_x = b \cos [\omega t - k(y \sin \theta + z \cos \theta)],$$

and

$$\mathcal{E}_x = b \exp [-jk(y \sin \theta + z \cos \theta)].$$

We know from the boundary conditions that the tangential component of **E**

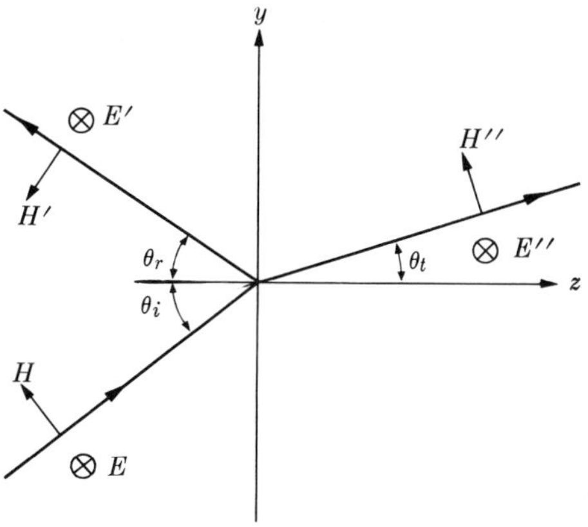

FIG. 9–3. Reflection and transmission at an interface.

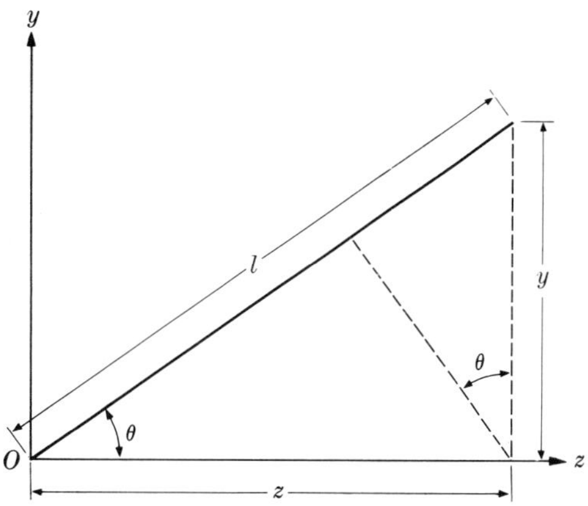

FIG. 9–4. Geometry for an oblique wave.

(and also of $\mathcal{E}$) must be continuous across the boundary, and thus, referring to Fig. 9-3, for $z = 0$ we have

$$E_x + E'_x = E''_x,$$

and so

$$\hat{E}_x \exp(-jk_1 y \sin \theta_i) + \hat{E}'_x \exp(-jk_1 y \sin \theta_r) = \hat{E}''_x \exp(-jk_2 y \sin \theta_t),$$

where $\hat{E}_x$ represents the peak value of the incident wave, etc. Since this must be true for all values of y, the variation with y of each of the three terms in this expression must be the same. Hence we have

$$k_1 \sin \theta_i = k_1 \sin \theta_r = k_2 \sin \theta_t.$$

This leads to $\theta_i = \theta_r$, that is, the angle of reflection is equal to the angle of incidence, a law we are familiar with in optics, and

$$\frac{\sin \theta_i}{\sin \theta_t} = \frac{k_2}{k_1} = \sqrt{\frac{\mu_2 \epsilon_2}{\mu_1 \epsilon_1}} = \frac{v_1}{v_2}.$$

The phase velocity $v = c/n$, where c is the velocity in vacuum and n is the refractive index. Hence we have

$$\frac{\sin \theta_i}{\sin \theta_t} = \frac{n_2}{n_1},$$

which is the familiar form of Snell's law in optics. Thus the electromagnetic theory of waves leads to the laws of reflection and refraction, which were discovered experimentally, and so we have another experimental check on the theory.

From the theory we can also deduce the magnitude of the reflection and transmission coefficients. Let E, E', and E'' represent the amplitude of the waves at $z = 0$. The boundary conditions for $\mathbf{E}$ require that $E + E' = E''$. We know that the tangential component of $\mathbf{H}$ must be continuous, and hence, from Fig. 9-3,

$$H \cos \theta_i - H' \cos \theta_i = H'' \cos \theta_t.$$

From the properties of the wave we also have

$$E/H = \sqrt{\mu/\epsilon}.$$

Let us consider the case of a nonmagnetic dielectric, so that

$$E/H = \sqrt{\mu_0/\epsilon} = \sqrt{\mu_0/\epsilon_0}\sqrt{\epsilon_0/\epsilon} = \frac{1}{n}\sqrt{\mu_0/\epsilon_0},$$

$$\therefore \frac{nE}{H} = \text{const.}$$

Using this result, we have the two equations

$$E + E' = E'', \qquad n_1 E \cos\theta_i - n_1 E' \cos\theta_i = n_2 E'' \cos\theta_t.$$

Eliminating E'' gives the reflection coefficient R:

$$R = \frac{E'}{E} = \frac{n_1 \cos\theta_i - n_2 \cos\theta_t}{n_1 \cos\theta_i + n_2 \cos\theta_t},$$

and the elimination of E' yields the transmission coefficient T:

$$T = \frac{E''}{E} = \frac{2n_1 \cos\theta_i}{n_1 \cos\theta_i + n_2 \cos\theta_t}.$$

These expressions were first derived by Fresnel and are known as Fresnel coefficients. The second pair of coefficients for the case where **H** is parallel to the interface will now be derived (Fig. 9–5). The boundary conditions for **H** will lead to the laws of reflection and refraction. For the Fresnel coefficients we proceed along similar lines as before:

$$H + H' = H'',$$

$$-E \cos\theta_i + E' \cos\theta_i = -E'' \cos\theta_t,$$

and using $(nE/H) = \text{const}$, we have

$$n_1 E + n_1 E' = n_2 E''.$$

By eliminating E'' we obtain the reflection coefficient

$$R = \frac{E'}{E} = \frac{n_2 \cos\theta_i - n_1 \cos\theta_t}{n_2 \cos\theta_i + n_1 \cos\theta_t},$$

and eliminating E' gives

$$T = \frac{E''}{E} = \frac{2n_1 \cos\theta_i}{n_2 \cos\theta_i + n_1 \cos\theta_t}.$$

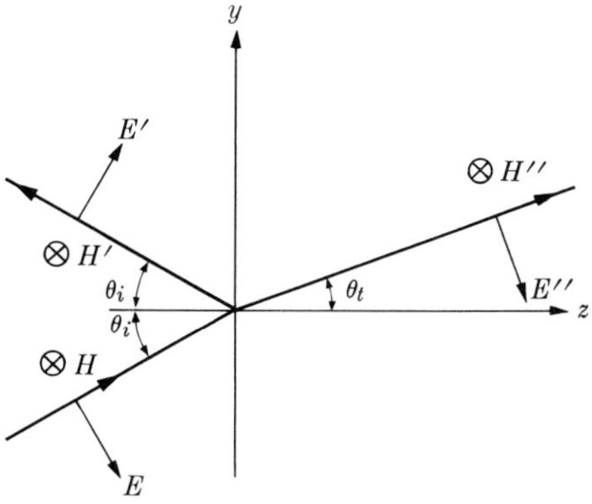

FIG. 9–5. Reflection and transmission at an interface.

These relations predict a phenomenon observed by Brewster. We can write the reflection coefficient, using $n_1 \sin\theta_i = n_2 \sin\theta_t$, as

$$R = \frac{\sin\theta_i \cos\theta_i - \sin\theta_t \cos\theta_t}{\sin\theta_i \cos\theta_i + \sin\theta_t \cos\theta_t} = \frac{\sin(\theta_i - \theta_t)\cos(\theta_i + \theta_t)}{\cos(\theta_i - \theta_t)\sin(\theta_i + \theta_t)},$$

$$\therefore R = \frac{\tan(\theta_i - \theta_t)}{\tan(\theta_i + \theta_t)}.$$

When $\theta_i + \theta_t = \pi/2$, $\tan(\theta_i + \theta_t)$ becomes infinite and $R = 0$. For this condition

$$\frac{n_2}{n_1} = \frac{\sin\theta_i}{\sin[(\pi/2) - \theta_i]} = \tan\theta_i.$$

This particular angle of incidence, $\tan^{-1}(n_2/n_1)$, is known as the Brewster angle. If a wave is incident at this angle, then the reflected wave is linearly polarized with the **E**-vector parallel to the interface regardless of the polarization of the incident wave.

Another special case of interest is a wave incident upon a good conductor, such as a metal. For simplicity we consider normal incidence. If we suppose that the first medium is vacuum, then we have

$$E/H = \sqrt{\mu_0/\epsilon_0},$$

and thus $n_1 = 1$ in the formula above. From the previous section we have, for a conductor,

$$E/H = \sqrt{j\omega\mu/(\sigma + j\omega\epsilon)}.$$

Assuming a nonmagnetic metal and a good conductor, so that $\mu = \mu_0$ and $\sigma \gg j\omega\epsilon$, we have

$$E/H = \sqrt{j\omega\epsilon_0/\sigma}\,\sqrt{\mu_0/\epsilon_0},$$

and therefore

$$n_2 = \sqrt{\sigma/j\omega\epsilon_0} = (1 - j)\sqrt{\sigma/2\omega\epsilon_0}.$$

Hence setting $\theta_i = \theta_t = 0$ we obtain, for the reflection coefficient,

$$R = \frac{n_2 - n_1}{n_2 + n_1} = \frac{(1 - j)\sqrt{\sigma/2\omega\epsilon_0} - 1}{(1 - j)\sqrt{\sigma/2\omega\epsilon_0} + 1}.$$

If $\sqrt{\sigma/2\omega\epsilon_0} \gg 1$, then, to a first approximation, $R = 1$, and the wave is completely reflected, in accordance with the fact that metals are good reflectors. To a better approximation this formula can be simplified to give

$$|R|^2 = 1 - \frac{2}{\sqrt{\sigma/2\omega\epsilon_0}},$$

which has also been verified experimentally by investigations in the infrared.

9-8 Radiation from a dipole

Let us consider a dipole consisting of two equal and opposite charges separated by a distance dl (Fig. 9-6). These, as we have seen, constitute a dipole of moment $P = Q\,dl$. We suppose that the charges vary sinusoidally with time, which means that charge flows from one end of the dipole to the other, since by the law of conservation of charge, we cannot create or destroy charges. This charge flow is equivalent to a current, which we can obtain as follows:

$$I = \frac{dQ}{dt}, \qquad P = Q\,dl,$$

$$\therefore I\,dl = \frac{dP}{dt}.$$

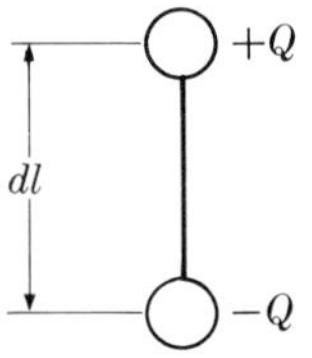

FIG. 9-6. A dipole.

If P has a simple harmonic variation with time, we can introduce complex quantities and write $\mathcal{I}\, dl = j\omega\mathcal{P}$. Thus an oscillating dipole (simple harmonic time variation) is equivalent to a time-harmonic current element ($\mathcal{I}\, dl$).

We saw earlier (Section 6–6) how to calculate the field due to an electric current distribution in the static case by using the magnetic vector potential. We shall now use a similar procedure to calculate the field for the current element varying sinusoidally with time. In general, the current density **J** is a function of position (a vector field). Here we have a very simple distribution since the current density is zero everywhere except at the one point occupied by the dipole. We choose this point as the origin of our coordinate system, and here the current density actually has an infinity in it and is such that a finite current I can flow in a filament of infinitesimal cross section. We will suppose this current to be flowing in the z-direction for an infinitesimal length of the z-axis dl. The time-harmonic equations which we have to solve are (Section 9–2)

$$-\text{curl}\, \mathcal{E} = j\omega\mu\mathcal{H},$$

$$\text{curl}\, \mathcal{H} = j\omega\epsilon\mathcal{E} + \mathcal{J},$$

where $\mathcal{J}$ is the distribution we have just deduced. Since div $\mathcal{B} = 0$, we can write, as in Section 6–6, $\mathcal{B} = \text{curl}\, \mathcal{A}$, except that the quantities now are complex and the fields themselves time-harmonic. Hence we have

$$-\text{curl}\, \mathcal{E} = j\omega\, \text{curl}\, \mathcal{A}, \qquad \text{or} \qquad \text{curl}\, (\mathcal{E} + j\omega\mathcal{A}) = 0.$$

We also know that a field with zero curl can be derived as the gradient of a scalar field, and thus we can write

$$\mathcal{E} + j\omega\mathcal{A} = -\text{grad}\, \phi.$$

(A minus sign is conventionally taken here, and this makes it consistent with the electrostatic case, since, if we make $\omega = 0$ (that is, zero frequency), we have $\mathcal{E} = -\text{grad}\, \phi$.) Thus

$$\mathcal{E} = -(j\omega\mathcal{A} + \text{grad}\, \phi).$$

From the second equation, we obtain

$$(1/\mu)\, \text{curl curl}\, \mathcal{A} = j\omega\epsilon\mathcal{E} + \mathcal{J},$$

and substituting for $\mathcal{E}$, we have

$$(1/\mu)\, \text{curl curl}\, \mathcal{A} = -j\omega\epsilon(j\omega\mathcal{A} + \text{grad}\, \phi) + \mathcal{J}.$$

Expanding curl curl, we arrive at

$$\text{grad div}\, \mathcal{A} - \nabla^2\mathcal{A} = \omega^2\mu\epsilon\mathcal{A} - j\omega\mu\epsilon\, \text{grad}\, \phi + \mu\mathcal{J}.$$

Thus far we have only defined the field $\mathcal{A}$ by the equation $\mathcal{B} = \operatorname{curl} \mathcal{A}$; that is, we have only specified the curl of the field. This does not completely specify $\mathcal{A}$ because, since curl grad $= 0$, we can add the gradient of any scalar field to $\mathcal{A}$ without altering the curl of the resulting field, i.e., curl $(\mathcal{A} + \operatorname{grad} w) = \mathcal{B}$. We can now make the divergence of the resulting field equal to anything we choose by suitably adjusting w. Thus, if we want the divergence of the field to be any function $f(x, y, z)$, then we would determine w from the equation

$$\operatorname{div} \mathcal{A} + \operatorname{div} \operatorname{grad} w = f(x, y, z).$$

Hence we see that we can specify the curl and the divergence of a vector field independently.

Reverting to the equation

$$\operatorname{grad} \operatorname{div} \mathcal{A} - \nabla^2 \mathcal{A} = \omega^2 \mu\epsilon \mathcal{A} - j\omega\mu\epsilon \operatorname{grad} \phi + \mu \mathcal{J},$$

in which div $\mathcal{A}$ is as yet unspecified, we see that this can be simplified by defining div $\mathcal{A}$ by the equation

$$\operatorname{div} \mathcal{A} = -j\omega\mu\epsilon\phi,$$

and substituting $\omega^2\mu\epsilon = k^2$, we have

$$(\nabla^2 + k^2)\mathcal{A} = -\mu\mathcal{J}.$$

From Section 6-6 we know that this must be interpreted to mean

$$(\nabla^2 + k^2)\mathcal{A}_x = -\mu\mathcal{J}_x, \qquad (\nabla^2 + k^2)\mathcal{A}_y = -\mu\mathcal{J}_y, \qquad (\nabla^2 + k^2)\mathcal{A}_z = -\mu\mathcal{J}_z.$$

In this case the only current present flows in the z-direction, and hence $\mathcal{J}_x = \mathcal{J}_y = 0$, giving the solutions $\mathcal{A}_x = \mathcal{A}_y = 0$. Although it is immediately clear that this result satisfies the equations, it is by no means obvious that this is the correct solution, and thus we give the following justification. We saw in Section 6-6 that for the static case

$$\mathbf{A} = \iiint \frac{\mu \mathbf{J}\, d\tau}{4\pi l}.$$

Hence for a single current element directed in the z-direction, the vector potential will also be in the z-direction. The addition of time variation will not affect the spatial direction, and therefore we would expect $\mathcal{A}$ to be in the z-direction, that is, $\mathcal{A}_x = \mathcal{A}_y = 0$. We thus have the equation $(\nabla^2 + k^2)\mathcal{A}_z = -\mu\mathcal{J}_z$ to solve. Note that both $\mathcal{A}_z$ and $\mathcal{J}_z$ are scalars, and the mathematical equation we are solving has the form

$$\nabla^2\psi + k^2\psi = -\mu g(x, y, z);$$

the function $g(x, y, z)$ is zero everywhere except at the origin where it has a pole. This is analogous to the point charge in electrostatics, and hence we would expect ψ to depend only upon the distance from the origin (cf. the potential due to a point charge), that is, $\psi = \psi(r)$. We have dealt with the operator ∇^2 only in Cartesian components thus far. It is clearly possible to express it in any coordinate system, and since we wish to deal in terms of the distance from the origin, spherical polars are appropriate. The expression for ∇^2 in spherical polars is, in general, rather complicated (Appendix A-5), but here it is only a function of r so that $\partial\psi/\partial\theta$ and $\partial\psi/\partial\phi$ are zero, and we have

$$\nabla^2\psi = \frac{1}{r^2}\frac{d}{dr}\left(r^2\frac{d\psi}{dr}\right) = \frac{1}{r}\frac{d^2}{dr^2}(r\psi).$$

Hence for all points except the origin we have

$$\frac{1}{r}\frac{d^2}{dr^2}(r\psi) + k^2\psi = 0, \qquad \text{or} \qquad \frac{d^2}{dr^2}(r\psi) = -k^2(r\psi).$$

This is the simple harmonic equation and has solutions

$$r\psi = Ce^{\pm jkr}.$$

From Section 9–3 we know that a variation like e^{-jkr} represents a wave traveling in the $+r$-direction, and e^{jkr} one in the $-r$-direction. Since we would expect the waves to travel outward from the source, we take the solution

$$\psi = \frac{C}{r}e^{-jkr},$$

where C is a constant to be determined. We obtain the constant of integration in solving a differential equation by taking a particular case for which the solution is known (e.g., in calculating the transient response of an RC-circuit we get the constant by taking $t = 0$). This is the equation for the magnitude of the vector potential due to a current element. We deduced the value of this potential for the case of a steady current in Section 6–7. There we obtained the expression

$$A_z = \iiint \frac{\mu J_z\, d\tau}{4\pi r}.$$

Since $J_z = 0$ except at the origin and $\iint J_z\, dx\, dy = I$, we have

$$A_z = \frac{\mu I\, dl}{4\pi r}.$$

But this is $\mathcal{A}_z$ for the particular case $k = 0$. Hence setting $k = 0$ in the expression above, we have an equation for C:

$$\mathcal{A}_z = \frac{\mu \mathcal{I}\, dl}{4\pi r} = \frac{C}{r},$$

$$\therefore C = \frac{\mu \mathcal{I}\, dl}{4\pi}.$$

Hence for $\mathcal{A}_z$ we finally obtain

$$\mathcal{A}_z = \frac{\mu \mathcal{I}\, dl}{4\pi}\frac{e^{-jkr}}{r}, \qquad \text{or} \qquad \boldsymbol{\mathcal{A}} = \frac{\mu \mathcal{I}\, dl}{4\pi}\frac{e^{-jkr}}{r}\mathbf{u}_z.$$

To complete the problem we obtain the $\boldsymbol{\mathcal{E}}$- and $\boldsymbol{\mathcal{H}}$-fields from $\boldsymbol{\mathcal{A}}$. We obtain $\boldsymbol{\mathcal{H}}$ immediately from the relation

$$\boldsymbol{\mathcal{H}} = (1/\mu)\,\mathrm{curl}\,\boldsymbol{\mathcal{A}}.$$

We also had

$$\boldsymbol{\mathcal{E}} = -\mathrm{grad}\,\phi - j\omega\boldsymbol{\mathcal{A}}, \qquad \text{and} \qquad \phi = -\frac{1}{j\omega\mu\epsilon}\,\mathrm{div}\,\boldsymbol{\mathcal{A}},$$

$$\therefore\ \boldsymbol{\mathcal{E}} = \frac{1}{j\omega\mu\epsilon}\,\mathrm{grad}\,\mathrm{div}\,\boldsymbol{\mathcal{A}} - j\omega\boldsymbol{\mathcal{A}}.$$

We note that the expression for $\boldsymbol{\mathcal{A}}$ is in mixed coordinates and the z-component of $\boldsymbol{\mathcal{A}}$ is expressed as a function of r. It is most convenient to carry through these calculations in sphericals. From the Appendix we have (remembering that $\mathcal{A}_x = \mathcal{A}_y = 0$)

$$\mathcal{A}_r = \mathcal{A}_z \cos\theta = \frac{\mu \mathcal{I}\, dl}{4\pi}\frac{e^{-jkr}}{r}\cos\theta,$$

$$\mathcal{A}_\theta = -\mathcal{A}_z \sin\theta = -\frac{\mu \mathcal{I}\, dl}{4\pi}\frac{e^{-jkr}}{r}\sin\theta,$$

$$\mathcal{A}_\phi = 0.$$

Using the expression for curl, we get

$$\mathcal{H}_r = 0, \qquad \mathcal{H}_\theta = 0,$$

$$\mathcal{H}_\phi = \frac{1}{\mu r}\left[\frac{\partial}{\partial r}(r\mathcal{A}_\theta) - \frac{\partial \mathcal{A}_r}{\partial \theta}\right] = \frac{\mathcal{I}\, dl}{4\pi}\,e^{-jkr}\left(\frac{jk}{r} + \frac{1}{r^2}\right)\sin\theta.$$

Using the relation for div $\boldsymbol{\mathcal{A}}$ in spherical polars, we find

$$\mathrm{div}\,\boldsymbol{\mathcal{A}} = -\frac{\mu \mathcal{I}\, dl}{4\pi}\,e^{-jkr}\left(\frac{jk}{r} + \frac{1}{r^2}\right)\cos\theta,$$

and from the formula

$$\boldsymbol{\mathcal{E}} = \frac{1}{j\omega\mu\epsilon}\,\mathrm{grad}\,\mathrm{div}\,\boldsymbol{\mathcal{A}} - j\omega\boldsymbol{\mathcal{A}},$$

we obtain

$$\mathcal{E}_r = \frac{\mathcal{I}\, dl}{2\pi}\,e^{-jkr}\left(\frac{\eta}{r^2} + \frac{1}{j\omega\epsilon r^3}\right)\cos\theta,$$

$$\mathcal{E}_\theta = \frac{\mathcal{I}\, dl}{4\pi}\,e^{-jkr}\left(\frac{j\omega\mu}{r} + \frac{\eta}{r^2} + \frac{1}{j\omega\epsilon r^3}\right)\sin\theta,$$

$$\mathcal{E}_\phi = 0,$$

where η is the intrinsic impedance of the medium. Note that $\mathcal{E}$ and $\mathcal{H}$ are not functions of ϕ. This means that the resulting field is symmetrical about the z-axis, which is what we would expect from physical considerations.

When r is very small, i.e., close to the radiator, the term with the largest power of r in the denominator will dominate the expressions and $e^{-jkr} \simeq 1$, so that for the fields we have

$$\mathcal{E}_r = \frac{\mathcal{I}\,dl}{2\pi}\frac{\cos\theta}{j\omega\epsilon r^3}, \qquad \mathcal{E}_\theta = \frac{\mathcal{I}\,dl}{4\pi}\frac{\sin\theta}{j\omega\epsilon r^3}, \qquad \mathcal{H}_\phi = \frac{\mathcal{I}\,dl}{4\pi}\frac{\sin\theta}{r^2}.$$

Since $\mathcal{I}\,dl = j\omega\mathcal{P}$, it follows that

$$\mathcal{E} = \frac{\mathcal{P}}{4\pi\epsilon r^3}(2\cos\theta\,\mathbf{u}_r + \sin\theta\,\mathbf{u}_\theta).$$

This is precisely the expression we obtained in Section 6-5 for the static dipole. Similarly, the expression for $\mathcal{H}$ is exactly that for the static current element obtained in Section 6-8. These fields are therefore referred to as *quasistatic*. Note that the fields **E** and **H** obtained from the above complex ones will be sinusoidally varying with time and that there is no restriction on the frequency.

At great distances from the radiator the $1/r$ term predominates, and we have

$$\mathcal{E}_\theta = \frac{\mathcal{I}\,dl\,j\omega\mu}{4\pi}\frac{e^{-jkr}}{r}\sin\theta,$$

$$\mathcal{H}_\phi = \frac{\mathcal{I}\,dl\,jk}{4\pi}\frac{e^{-jkr}}{r}\sin\theta,$$

which is referred to as the radiation field. Note that it is a purely transverse field—there is no component of **E** or **H** in the direction of propagation. Also the directions of **E** and **H** and the direction of propagation form a right-handed orthogonal set just as in the case of the plane wave. The impedance is

$$\frac{E_\theta}{H_\phi} = \frac{\omega\mu}{k} = \sqrt{\mu/\epsilon} = \eta,$$

as for the plane wave. If we convert to the **E**- and **H**-fields, we see that these vary like $\cos(\omega t - kr)$. Hence for constant r all the oscillators are in phase, and the surface of equal phase angle is therefore $r =$ const. But this is a sphere, and thus the dipole radiates spherical waves. The magnitude of these fields is a function of θ, and so will not be constant over the sphere. Hence the wave is nonuniform. It is this variation of the radiated amplitude with direction that leads to the polar diagram of an antenna. For the radiation field the electric vector has only one component, namely, a θ-component. At any one point in space, the electric vector always points in the θ-direction, which is a fixed direction at a fixed point. As time varies the magnitude of this vector increases and decreases, but its direction remains constant. The tip of the vector must therefore trace a line, and the wave is linearly polarized.

9-9 The solution for general time variation

In this chapter we have considered fields which vary sinusoidally with time. This is of great importance in electrical engineering, but the case of general time variation is also of interest. Considering a dielectric, we have, from Maxwell's equations,

$$-\text{curl}\,\mathbf{E} = \mu \frac{\partial \mathbf{H}}{\partial t}, \qquad \text{curl}\,\mathbf{H} = \epsilon \frac{\partial \mathbf{E}}{\partial t}.$$

By eliminating **H**, we obtain

$$-\text{curl curl}\,\mathbf{E} = \mu \frac{\partial}{\partial t}\,\text{curl}\,\mathbf{H} = \mu\epsilon \frac{\partial^2 \mathbf{E}}{\partial t^2}.$$

Expanding curl curl **E** and assuming div **E** $= 0$ (no sources), we have

$$\nabla^2 \mathbf{E} = \mu\epsilon \frac{\partial^2 \mathbf{E}}{\partial t^2}.$$

This is the classic form of the wave equation. Making assumptions similar to those made for the time-harmonic case, we will consider the situation where **E** has only an x-component, which is a function of z only. The equation then reduces to

$$\frac{\partial^2 E_x}{\partial z^2} = \mu\epsilon \frac{\partial^2 E_x}{\partial t^2}.$$

It is readily verified that the functions $f(t - \sqrt{\mu\epsilon}\, z)$ and $g(t + \sqrt{\mu\epsilon}\, z)$ satisfy this equation, so that the general solution is

$$E_x = a \quad f(t - \sqrt{\mu\epsilon}\, z) + b \quad g(t + \sqrt{\mu\epsilon}\, z),$$

where a and b are arbitrary constants and f and g arbitrary functions. We can use the same argument as in Section 9-3 to understand the nature of the

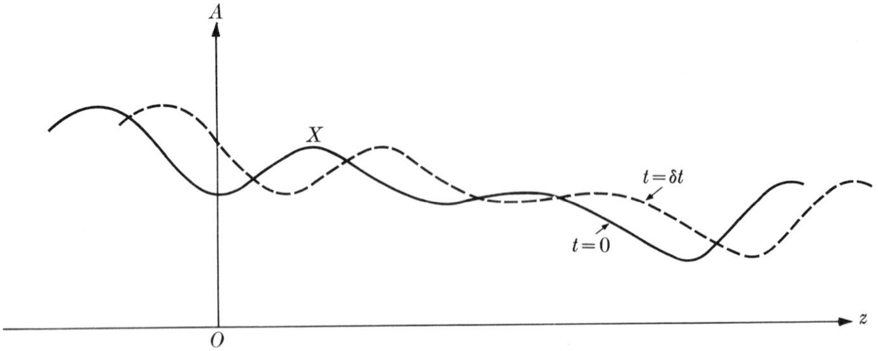

FIG. 9-7. A traveling wave.

function $F(t, z) = f(t - \sqrt{\mu\epsilon}\, z)$. At $t = 0$ we have $F = f(-\sqrt{\mu\epsilon}\, z)$, and we can plot a graph of F against z (Fig. 9–7). Choosing any feature on this arbitrary curve, such as the trough shown at the origin or the crest at X, we find that at time δt this will have moved a distance δz to the right, where

$$\delta t - \sqrt{\mu\epsilon}\, \delta z = 0,$$

that is, in time δt it moves a distance δz, and thus the velocity is given by

$$v = \frac{\delta z}{\partial t} = \frac{1}{\sqrt{\mu\epsilon}}.$$

Hence the general case allows us to have any arbitrary curve for amplitude plotted as a function of z, and as time progresses this curve moves as a whole.

Problems

9–1 If div $\mathbf{D} = 0$, show that div $\mathfrak{D} = 0$.

9–2 Prove that curl [Re $(\boldsymbol{\mathcal{E}} e^{j\omega t})$] = Re [(curl $\boldsymbol{\mathcal{E}})e^{j\omega t}$].

9–3 Show that the boundary conditions for $\boldsymbol{\mathcal{E}}$ are the same as those for $\mathbf{E}$.

9–4 Show that $\mathcal{E}_z = Ae^{jkz} + Be^{-jkz}$ is not a possible solution to the equations developed in Section 9–3 for waves in a dielectric. [*Hint:* Consider the divergence of the fields.]

9–5 Derive the appropriate solutions to the equations of Section 9–3, given that the wave is to be circularly polarized. Derive them for (a) right-handed and (b) left-handed polarization.

9–6 For the solution $\mathcal{E}_x = Ae^{jkz} + Be^{-jkz}$, the wave is a mixture of a traveling and a standing wave. Calculate the Poynting vector, assuming $B > A$.

9–7 Calculate the phase velocity for the wave of Problem 9–6. By taking a limiting case obtain the phase velocity of a pure standing wave.

9–8 A loss-free medium has a linearly polarized uniform plane wave of magnitude A traveling in the $+z$-direction, and one of magnitude B in the $-z$-direction. Sketch the variation of $\mathbf{E}$ with z for values of ωt of 0, $\pi/2$, π, $3\pi/2$, 2π. If an observer moves with the crest of the wave, what variation in amplitude does he see?

9–9 An absorbing medium contains two waves traveling in opposite directions. Describe qualitatively how the standing wave ratio will vary.

9–10 A uniform, plane, linearly polarized wave traveling in a medium with dielectric constant κ_1 is incident normally on a plane boundary, and the dielectric constant on the other side of the boundary is κ_2. Starting from the boundary conditions, calculate (a) the reflection coefficient, (b) the transmission coefficient, and (c) the standing-wave ratio in the first medium.

9–11 Using the results of Section 9–7, show that for metals (which have very high conductivity) the reflection coefficient R for a wave incident normally from a vacuum can be written

$$|R|^2 = 1 - \frac{2}{\sqrt{(\sigma/2\omega\epsilon_0)}}.$$

9–12 Calculate the magnitude of the reflection coefficient of Problem 9–11 for copper ($\sigma = 5 \times 10^7$ mho/m) at 1 Mc/sec and at 1000 Mc/sec.

9–13 Calculate the Poynting vector (time-dependent form) for the field radiated by a dipole. Note the direction of the zero and the maximum of the radiated power density relative to the orientation of the dipole.

CHAPTER 10

The Theory of Microwaves

10–1 Introduction

Waveguides are now commonplace in electrical engineering; they form the "wires" at the higher frequencies. When we connect two points by a wire in an electrical circuit, what we are basically doing is providing a means for transferring electrical energy between the two points. To do this efficiently there must be no energy loss on the way. We use copper wires for this purpose, since copper is a good conductor and thus dissipates little energy in the form of heat. When the length of the wires becomes of the same order of magnitude as the wavelength we are dealing with, then the simple ideas of circuit theory are no longer adequate, and we must use the concepts of transmission line theory. As wavelengths get shorter and frequencies higher, we eventually reach a stage when a hollow conductor or waveguide can be used to guide the energy. The advantage of the waveguide is that the losses are smaller. We may ask why they are not used at all frequencies. Our studies in this chapter will provide the answers to such questions. We shall see that a waveguide has a cutoff frequency below which energy cannot flow down it. This depends upon the physical dimensions of the guide, so that, although it is theoretically possible to have waveguides at lower frequencies, they become impractical below about 300 Mc/sec.

10–2 The rectangular waveguide

In this section we deduce a solution of Maxwell's equations which satisfies the conditions imposed by a hollow rectangular conductor of infinite length (Fig. 10–1). We shall use the idea of sinusoidal time variation and complex fields introduced in the last chapter. The equations for a lossless medium can be written

$$-\text{curl}\,\boldsymbol{\mathcal{E}} = j\omega\mu\boldsymbol{\mathcal{H}}, \qquad \text{curl}\,\boldsymbol{\mathcal{H}} = j\omega\epsilon\boldsymbol{\mathcal{E}}.$$

Eliminating $\boldsymbol{\mathcal{H}}$ as before, we have

$$(\nabla^2 + k^2)\boldsymbol{\mathcal{E}} = 0,$$

meaning that each of the Cartesian components of $\boldsymbol{\mathcal{E}}$ satisfies this equation. Let us consider the z-component

$$(\nabla^2 + k^2)\mathcal{E}_z = 0.$$

To solve this equation we must express ∇^2 in terms of some coordinate system, and we choose the Cartesian. Hence we have

$$\frac{\partial^2 \mathcal{E}_z}{\partial x^2} + \frac{\partial^2 \mathcal{E}_z}{\partial y^2} + \frac{\partial^2 \mathcal{E}_z}{\partial z^2} + k^2 \mathcal{E}_z = 0.$$

In a similar way, by eliminating $\boldsymbol{\mathcal{E}}$ from the original equations, we may show that

$$\frac{\partial^2 \mathcal{H}_z}{\partial x^2} + \frac{\partial^2 \mathcal{H}_z}{\partial y^2} + \frac{\partial^2 \mathcal{H}_z}{\partial z^2} + k^2 \mathcal{H}_z = 0.$$

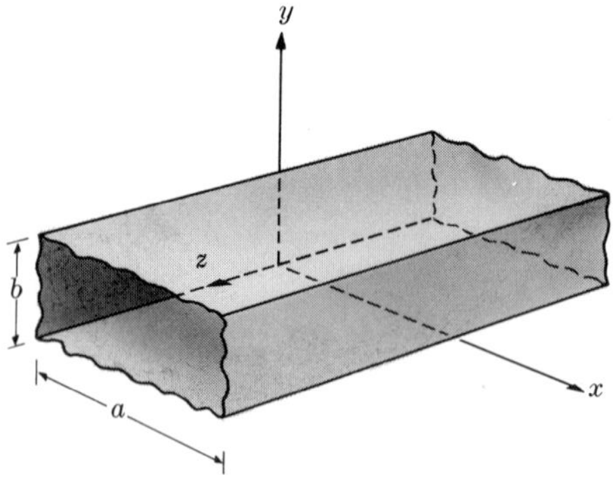

FIG. 10–1. A rectangular waveguide.

The standard method of solving this type of equation is by separation of variables. In this process we use the trial solution

$$\mathcal{E}_z = X(x)Y(y)Z(z),$$

that is, we assume that the solution can be expressed as the product of three terms, each of which depends on only one of the three variables. Substituting this solution and dividing by $\mathcal{E}_z$ gives

$$\frac{1}{X}\frac{d^2X}{dx^2} + \frac{1}{Y}\frac{d^2Y}{dy^2} + \frac{1}{Z}\frac{d^2Z}{dz^2} + k^2 = 0.$$

Since k is a constant (i.e., independent of x, y, and z), the sum of the first three terms is constant. But x, y, and z can vary independently; thus, if we hold y and z constant, the second and third terms are constants, and therefore the first term must also be constant regardless of any variation in x. A similar argument applies to the other two terms, and we can write

$$\frac{d^2X}{dx^2} = -k_x^2 X, \qquad \frac{d^2Y}{dy^2} = -k_y^2 Y, \qquad \frac{d^2Z}{dz^2} = -k_z^2 Z,$$

where the constants must obey the *separation equation*

$$k_x^2 + k_y^2 + k_z^2 = k^2.$$

The expressions for X, Y, and Z are just the simple harmonic-motion equations. From our work on waves in Chapter 9 we know that energy is carried by a traveling wave in the direction in which the wave is traveling. If we are to have energy carried down the guide, i.e., in the z-direction, we need a wave traveling in the z-direction. From the last chapter we know that a solution of the form $e^{-jk_z z}$ will give such a traveling wave, and we also see from the equation for Z that such a solution is possible for this case. Hence we have

$$\mathcal{E}_z = X(x)Y(y)e^{-jk_z z}.$$

We shall investigate the restrictions imposed on the solution by the waveguide. If we assume that the walls of the guide are perfect conductors, then the tangential components of the electric field at the walls of the guide must be zero. Since $\mathcal{E}_z$ is tangential to the walls of the guide, it must be zero at $x = 0$ and a, and $y = 0$ and b. The only function which is zero at the origin and satisfies the simple harmonic equation is the sine function, and thus we have

$$\mathcal{E}_z = \sin k_x x \sin k_y y \; e^{-jk_z z}.$$

Since $\mathcal{E}_z = 0$ also at $x = a$ and $y = b$, we obtain

$$k_x = \frac{m\pi}{a}, \qquad k_y = \frac{n\pi}{b},$$

where m and n are integers, and hence

$$\mathcal{E}_z = \sin \frac{m\pi}{a} x \sin \frac{n\pi}{b} y \; e^{-jk_z z},$$

and k_z is given by the equation

$$\left(\frac{m\pi}{a}\right)^2 + \left(\frac{n\pi}{b}\right)^2 + k_z^2 = k^2.$$

We see that there is a doubly infinite set of solutions for this case, since both m and n can go from unity to infinity. The particular values of k_x and k_y which are possible in a given problem are known as the *eigenvalues* for the problem, and the functions $\sin (m\pi/a)x$ and $\sin (n\pi/b)y$ are the *eigenfunctions*. We note that we now have an expression for $\mathcal{E}_z$ in terms of a and b (the dimensions of the guide), k (which contains the frequency and the permeability and permittivity of the medium), and m and n (which are integers). The numbers m and n are mode numbers, and the solutions are known as the *natural modes* of the guide. These modes are labeled the TM_{mn}-modes. Note that the lowest-order TM-mode is the TM_{11}-mode, since if either m or n is zero, then $\mathcal{E}_z$ is zero.

We might attempt to obtain the other components of this field in the same way. For example, $\mathcal{E}_x$ obeys the same equation as $\mathcal{E}_z$, and thus we would write

$$\mathcal{E}_x = X(x)Y(y)Z(z).$$

We take $Z(z) = e^{-jk_z z}$ to obtain a wave traveling in the z-direction, and X and Y obey the simple harmonic equations. We have already determined the eigenvalues for the problem, and hence it remains to choose simple harmonic functions for X and Y which satisfy the boundary conditions. Since $\mathcal{E}_x$ is a tangential component of the electric field at the wall of the guide $y = 0$ and $y = b$, then it must be zero at these points; hence $Y(y) = \sin (n\pi/b)y$. However, the boundary conditions do not enable us to determine X directly, and $\mathcal{E}_x$ cannot be determined completely by this method. Note, though, that the components of the field are all interrelated by the original curl equations.

The two curl equations are

$$\operatorname{curl} \boldsymbol{\mathcal{E}} = -j\omega\mu\boldsymbol{\mathcal{H}}, \qquad \operatorname{curl} \boldsymbol{\mathcal{H}} = j\omega\epsilon\boldsymbol{\mathcal{E}}.$$

Writing these in components, we have

$$\frac{\partial \mathcal{E}_z}{\partial y} - \frac{\partial \mathcal{E}_y}{\partial z} = -j\omega\mu\mathcal{H}_x, \qquad \frac{\partial \mathcal{H}_z}{\partial y} - \frac{\partial \mathcal{H}_y}{\partial z} = j\omega\epsilon\mathcal{E}_x,$$

$$\frac{\partial \mathcal{E}_x}{\partial z} - \frac{\partial \mathcal{E}_z}{\partial x} = -j\omega\mu\mathcal{H}_y, \qquad \frac{\partial \mathcal{H}_x}{\partial z} - \frac{\partial \mathcal{H}_z}{\partial x} = j\omega\epsilon\mathcal{E}_y,$$

$$\frac{\partial \mathcal{E}_y}{\partial x} - \frac{\partial \mathcal{E}_x}{\partial y} = -j\omega\mu\mathcal{H}_z, \qquad \frac{\partial \mathcal{H}_y}{\partial x} - \frac{\partial \mathcal{H}_x}{\partial y} = j\omega\epsilon\mathcal{E}_z.$$

If we suppose that the variation with z is $e^{-jk_z z}$, then $\partial/\partial z$ becomes $-jk_z$, and we have

$$\frac{\partial \mathcal{E}_z}{\partial y} + jk_z\mathcal{E}_y = -j\omega\mu\mathcal{H}_x, \qquad \frac{\partial \mathcal{H}_z}{\partial y} + jk_z\mathcal{H}_y = j\omega\epsilon\mathcal{E}_x,$$

$$-jk_z\mathcal{E}_x - \frac{\partial \mathcal{E}_z}{\partial x} = -j\omega\mu\mathcal{H}_y, \quad -jk_z\mathcal{H}_x - \frac{\partial \mathcal{H}_z}{\partial x} = j\omega\epsilon\mathcal{E}_y,$$

$$\frac{\partial \mathcal{E}_y}{\partial x} - \frac{\partial \mathcal{E}_x}{\partial y} = -j\omega\mu\mathcal{H}_z \qquad \frac{\partial \mathcal{H}_y}{\partial x} - \frac{\partial \mathcal{H}_x}{\partial y} = j\omega\epsilon\mathcal{E}_z.$$

Eliminating $\mathcal{H}_y$ between the second and fourth equations yields

$$-k_z^2\mathcal{E}_x + jk_z\frac{\partial \mathcal{E}_z}{\partial x} + j\omega\mu\frac{\partial \mathcal{H}_z}{\partial y} = -k^2\mathcal{E}_x,$$

$$\mathcal{E}_x = -\frac{1}{k^2 - k_z^2}\left(jk_z\frac{\partial \mathcal{E}_z}{\partial x} + j\omega\mu\frac{\partial \mathcal{H}_z}{\partial y}\right),$$

which gives us an expression for $\mathcal{E}_x$ in terms of the two longitudinal components $\mathcal{E}_z$ and $\mathcal{H}_z$. Similarly $\mathcal{E}_y$, $\mathcal{H}_x$, and $\mathcal{H}_y$ can be obtained in terms of these two quantities. The form of the equations shows that the problem may be split into two parts. We may consider $\mathcal{H}_z = 0$ (which satisfies the equations) and investigate the nonzero solutions for $\mathcal{E}_z$ and vice versa. The general solution will then consist of a superposition of the two. The first set of solutions for which $\mathcal{H}_z = 0$ has a magnetic field which is perpendicular to the direction of propagation, and these waves are known as transverse magnetic or TM-waves. The second set has $\mathcal{E}_z = 0$, and these are the transverse electric or TE-waves.

Returning to the TM-wave, we obtain $\mathcal{E}_x$ from the above equation as

$$\mathcal{E}_x = -\frac{jk_z}{k^2 - k_z^2}\frac{\partial \mathcal{E}_z}{\partial x}.$$

But we have already solved for $\mathcal{E}_z$ and hence

$$\mathcal{E}_x = -\frac{jk_z}{k^2 - k_z^2}\frac{m\pi}{a}\cos\frac{m\pi}{a}x\sin\frac{n\pi}{b}y\,e^{-jk_z z}.$$

Note that this has the same variation with y which we previously deduced from the boundary conditions. The other components of the TM-field can be obtained in a similar fashion.

To complete the investigation of the solution we must consider the TE-waves. We already have the expressions for the x- and y-components of $\boldsymbol{\mathcal{E}}$ and $\boldsymbol{\mathcal{H}}$ in terms of the z-components. For example, setting $\mathcal{E}_z = 0$ gives, for $\mathcal{E}_x$,

$$\mathcal{E}_x = -\frac{j\omega\mu}{k^2 - k_z^2}\frac{\partial \mathcal{H}_z}{\partial y},$$

and we must now determine $\mathcal{H}_z$. Following the same steps as we used for $\mathcal{E}_z$, we obtain

$$\mathcal{H}_z = X(x)Y(y)e^{-jk_z z},$$

where X and Y obey the simple harmonic equation. But we cannot continue in the same way since the boundary condition does not give us any information about $\boldsymbol{\mathcal{H}}$ directly. Since the walls of the guide are electric conductors, the boundary conditions place restrictions on the electric field. Hence we must derive the electric field from $\mathcal{H}_z$ and we obtain

$$\mathcal{E}_x = -\frac{j\omega\mu}{k^2 - k_z^2}X(x)Y'(y)e^{-jk_z z}.$$

Since $\mathcal{E}_x$ is the tangential component of the electric field at the walls of the guide $y = 0$ and $y = b$, it must be zero at these points. Hence $Y'(y)$ must be zero at $y = 0, b$, and thus we take Y' to be sinusoidal and $Y = \cos(n\pi/b)y$. Similarly, using $\mathcal{E}_y$ we deduce $X = \cos(m\pi/a)x$, and we have

$$\mathcal{H}_z = \cos\frac{m\pi}{a}x\cos\frac{n\pi}{b}y\,e^{-jk_z z},$$

$$\mathcal{E}_x = \frac{j\omega\mu}{k^2 - k_z^2}\frac{n\pi}{b}\cos\frac{m\pi}{a}x\sin\frac{n\pi}{b}y\,e^{-jk_z z},$$

and similar results for the other components. Note that we can have $m = 0$ or $n = 0$ without all components of the field collapsing to zero everywhere. We cannot, however, have both $m = 0$ and $n = 0$ without this happening.

It is important to note that although the curl equations provide six equations among six unknowns, the six unknowns cannot be determined uniquely, because we have a set of homogeneous equations, and it is readily verified that if we have one set of values for the unknowns which satisfy the equations, then multiplication of all the quantities by the same constant will yield values which also satisfy the equations. The six original equations are not, in fact, independent. Only five of them are independent, and thus we can solve only for the ratios of the unknowns, and the solution contains an arbitrary constant multiplying factor. This is exactly what we would expect from physical considerations. Since we are dealing with a linear system, the amplitude of the oscillations is arbitrary. In practice, of course, this is determined by the source driving the system.

10–3 Some properties of the waves

In solving the problems of the last section, three constants were introduced. Two of them, k_x and k_y, are determined by the physical dimensions of the guide, and the third, k_z, is obtained from the separation equation:

$$k_x^2 + k_y^2 + k_z^2 = k^2, \qquad k_z^2 = \omega^2\mu\epsilon - \left(\frac{m\pi}{a}\right)^2 - \left(\frac{n\pi}{b}\right)^2$$

We see that for a given waveguide and a given medium within it, all terms are constants except the frequency. Hence for frequencies which are high enough, k_z^2 will be positive, and for frequencies low enough, it will be negative. We shall look into the significance of this statement. We chose a variation with z of e^{-jk_zz} to obtain a wave traveling in the z-direction. We succeed in obtaining this type of wave only if k_z is real, i.e., at high enough frequencies. It is conventional to put $k_z = \beta$ for this case. There clearly is a particular value of the frequency which makes $k_z = 0$. This is known as the cutoff frequency for reasons which will become apparent shortly. The particular value of the wave number which this frequency gives is similarly known as the cutoff wave number. Thus we have

$$\omega_c^2\mu\epsilon = k_c^2 = \left(\frac{m\pi}{a}\right)^2 + \left(\frac{n\pi}{b}\right)^2.$$

We can now write

$$k_z^2 = k^2 - k_c^2,$$

and

$$\beta = \sqrt{k^2 - k_c^2}, \qquad k > k_c.$$

The variation of the field quantities with time and distance in the z-direction will be $\cos(\omega t - \beta z)$. Since the phase angle of this oscillation is $(-\beta z)$, it follows that β, known as the *phase constant*, represents minus the rate of change of phase angle with z. By arguments similar to those used in Section 9–3, we see that this wave is a plane wave, since $z =$ const defines a plane perpendicular to the z-direction, and for all points with the same value of z, all oscillations have the same phase angle. Again, following the arguments of Section 9–3, we see that the phase velocity is given by ω/β. Let us consider the uniformity of the wave. The surfaces of equal phase angle are planes parallel to the xy-plane, and hence for the wave to be uniform the amplitude of the oscillation would have to be independent of x and y. However, this is not the case. For example, E_x behaves like $\cos (m\pi/a)x \sin (n\pi/b)y$. Hence we have a nonuniform plane wave in the guide.

Note that in this case, β plays a role similar to that played by k in Section 9–3. It is important to distinguish between the two. The wave number k is a property of the medium we are dealing with, since it contains μ and ϵ, and of the frequency ω. The phase constant represents the rate of change of phase angle with z. The two are, however, related by the formula $\beta = \sqrt{k^2 - k_c^2}$. For the special case $k_c = 0$, we have $\beta = k$. We might think of this as arising by letting the cross section of the guide become infinitely large, i.e., by letting $a \to \infty$ and $b \to \infty$ which will make $k_c = 0$. It will also remove the variation of amplitude with x and y. This is readily seen for the term $\cos (m\pi/a)x$, which becomes unity as $a \to \infty$. The sine terms need a more careful investigation. For example, E_x for the TE-waves varies like $\sin n\pi y/b$, which at first sight would indicate that the solution would collapse to zero if we make $b \to \infty$. However, we note that E_x also contains the term $k^2 - k_z^2$ in the denominator, which is zero here, and the expression does, in fact, remain finite. We then have the result that for the special case of the uniform plane wave the wave number is equal to the phase constant. Stated in another way, the intrinsic phase constant is equal to the wave number. We can see that logically we should have been dealing with the phase constant in Section 9–3, but it happened that this turned out to be the same as the wave number, and thus the need for distinguishing between the two did not present itself at that stage.

To complete our description of the wave, we must investigate its polarization. We note that E_x and E_y vary differently with x and y, and hence the polarization is not the same at all points across the section of the guide. On the other hand, E_x and E_y vary in the same way with z, and hence the polarization does not change with z. To illustrate the point, if we consider the line down the center of the guide for example, that is, $x = a/2$, $y = b/2$, then the polarization of the wave is the same at all points on this line. The polarization therefore does not change as the wave propagates down the guide. This property of a wave is of importance when we consider more complicated types of media (see Chapter 12).

We must now consider the case where $\omega < \omega_c$. Here k_z is imaginary, and it is conventional to put $k_z = -j\alpha$. Hence we have

$$k_z^2 = -\alpha^2 = k^2 - k_c^2, \qquad \text{and} \qquad \alpha = \sqrt{k_c^2 - k^2}.$$

The variation of the complex fields with z becomes $e^{-jk_z z} = e^{-\alpha z}$, and the time-varying fields have a variation $e^{-\alpha z} \cos \omega t$. This no longer gives us a traveling wave, which was the basis on which we chose the function for the z-variation, but it is nevertheless a perfectly good mathematical solution to the equations, and we must inquire into its physical significance. We see immediately that the variation with time is $\cos \omega t$ at all points, i.e., the oscillations are everywhere in phase. The wave is consequently not a traveling wave. It has this property in common with the standing wave. However, unlike the standing wave, the amplitude of the oscillations falls off exponentially with z instead of being periodic. This wave belongs to a new class of waves which is termed *evanescent*. For frequencies where the waves are evanescent and no propagation can take place, the guide is said to be cut off, and the critical frequency at which the wave type changes from traveling to evanescent is accordingly known as the cutoff frequency. As we shall see later, no power is transferred by evanescent waves.

The equation derived above, $\beta = \sqrt{k^2 - k_c^2}$, provides a relation between the phase constant and the frequency for a given mode in a given waveguide. A plot of this gives what is commonly referred to as the $\omega\beta$-diagram, which is of great importance in studying the types of waves that can exist on more complicated structures such as corrugated waveguides or a helix. These structures have important applications in such diverse fields as microwave tubes and high-energy particle accelerators. For our simple case we have

$$\beta = \sqrt{\omega^2 \mu\epsilon - k_c^2}.$$

For large frequencies such that $\omega^2\mu\epsilon \gg k_c^2$, we have $\beta = \pm\omega\sqrt{\mu\epsilon}$, and the curve becomes asymptotic to these two straight lines. For $\beta = 0$, $\omega\sqrt{\mu\epsilon} = k_c$. The $\omega\beta$-diagram therefore takes on the form shown in Fig. 10–2.

The phase velocity at any frequency is readily obtained from this diagram as the slope of the line drawn from the origin to the point on the curve corresponding to that frequency. Thus at frequency ω_1 the corresponding point on the curve is A, and ω/β is given by AN/ON. The straight line $\omega = \beta/\sqrt{\mu\epsilon}$,

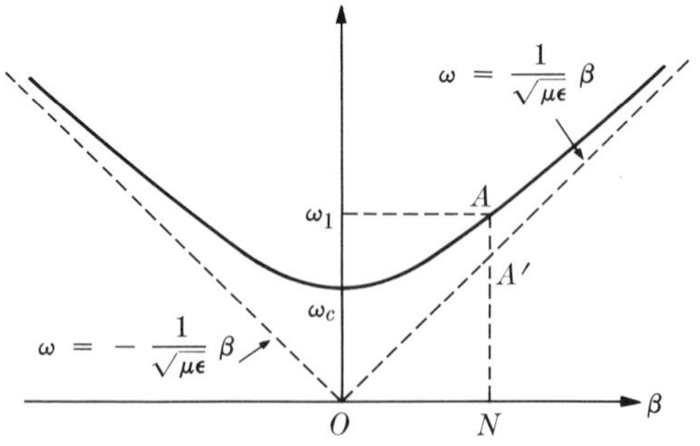

FIG. 10–2. The $\omega\beta$-diagram for a rectangular waveguide.

to which the curve is asymptotic at large frequencies, has a slope ω/k or $1/\sqrt{\mu\epsilon}$, which is the intrinsic velocity. Since the curve lies above the straight line so that $AN/ON > A'N/ON$, we see that the phase velocity in the guide is greater than the intrinsic velocity (often referred to loosely as the velocity of light).

The group velocity is also readily obtained from the diagram. This is a term which we have not yet defined. Let us consider an amplitude-modulated signal and suppose that the modulation frequency is small compared to the carrier frequency; then we obtain

$$\begin{aligned} f(t) &= (1 + m \cos \Delta\omega\, t) \cos \omega t \\ &= \cos \omega t + \tfrac{1}{2}m[\cos (\omega - \Delta\omega)t + \cos (\omega + \Delta\omega)t]. \end{aligned}$$

If β is a function of ω, we suppose that the first component in the brackets has a phase constant $\beta - \Delta\beta$ and the second $\beta + \Delta\beta$. Hence for a propagating disturbance we have

$$\begin{aligned} f(z, t) &= \cos (\omega t - \beta z) + \tfrac{1}{2}m\{\cos [(\omega - \Delta\omega)t - (\beta - \Delta\beta)z] \\ &\qquad\qquad + \cos [(\omega + \Delta\omega)t - (\beta + \Delta\beta)z]\} \\ &= [1 + m \cos (\Delta\omega\, t - \Delta\beta\, z)] \cos (\omega t - \beta z). \end{aligned}$$

Thus the modulation envelope propagates with a velocity $\Delta\omega/\Delta\beta$, which is defined as the group velocity (v_g). The speed at which the modulation propagates represents the speed at which we can signal. This is obtained from the $\omega\beta$-diagram as the slope of the curve; it is less than the slope of the asymptote, and thus the group velocity is less than the intrinsic velocity. At the cutoff frequency where $\beta = 0$ the slope of the curve is zero, and hence the group velocity is zero, which also fits in with the idea of the waveguide being cut off.

We may note in passing that a system in which the phase velocity (or phase constant) is a function of frequency is said to be *dispersive*. We note that

$$\frac{dv_p}{d\beta} = \frac{dv_p}{d\omega} \cdot \frac{d\omega}{d\beta} = v_g \frac{dv_p}{d\omega},$$

and since

$$v_p = \frac{\omega}{\beta}, \qquad \frac{dv_p}{d\beta} = \frac{1}{\beta}\frac{d\omega}{d\beta} - \frac{\omega}{\beta^2} = \frac{1}{\beta}(v_g - v_p),$$

it follows that

$$v_g = \frac{v_p}{1 - \beta(dv_p/d\omega)},$$

and if v_p is not a function of ω, $v_g = v_p$. For a nondispersive medium the phase and group velocities are equal. For most cases of dispersive media of interest to the electrical engineer, the group velocity may be thought of as above, and it is equal to the signal velocity. However, this is not invariably true. Some media exhibit what has been termed *anomalous dispersion.* Here the simple concept of group velocity as defined above loses its significance, and group and signal velocities are different. We shall not, however, pursue this matter further, but will return to a consideration of the cutoff frequency. We have

$$k_c^2 = \left(\frac{m\pi}{a}\right)^2 - \left(\frac{n\pi}{b}\right)^2,$$

and we see that the cutoff frequency depends on m and n. Thus each mode has a different cutoff frequency, and the lowest one is obtained for $a > b$ by making $m = 1$ and $n = 0$. From Section 10-2 we see that the lowest-order nonzero TM-mode is the TM_{11}-mode. Hence the mode with the lowest cutoff frequency must be a TE-mode, and is therefore the TE_{10}-mode. Note that if we had chosen to make $a < b$, then it would have been the TE_{01}-mode which had the lowest cutoff frequency. For the TE_{10}-mode cutoff, we have

$$k_c = \frac{\pi}{a}, \qquad \text{or} \qquad \omega_c = \frac{\pi}{a\sqrt{\mu\epsilon}}.$$

The mode with the lowest cutoff frequency is known as the *dominant mode.* If we make $a = 2b$, the next modes to become propagating are the TE_{01} and the TE_{20}. Both have a cutoff wave number of $2\pi/a$, which is twice that for the TE_{10}-wave. Hence we have a 2:1 range of frequencies over which only one mode will propagate; all the rest are evanescent. This is a fairly common practice. The waveguide is designed with its sides in the ratio 2:1 and their dimensions such that the working frequency is in the range where only the dominant mode can propagate. Note that in this example we have two modes with the same cutoff frequency. Such modes are said to be *degenerate.*

It is convenient to regard cutoff in terms of the intrinsic cutoff wavelength. For example, for the TE_{10}-mode, this is given by

$$\lambda_c = \frac{(v_p)_i}{f_c} = \frac{2\pi}{\omega_c\sqrt{\mu\epsilon}},$$

where $(v_p)_i$ is the intrinsic phase velocity. But we also have

$$\omega_c^2\mu\epsilon = \left(\frac{m\pi}{a}\right)^2 + \left(\frac{n\pi}{b}\right)^2,$$

which for the TE_{10}-mode gives

$$\omega_c\sqrt{\mu\epsilon} = \frac{\pi}{a}, \qquad \lambda_c = 2a,$$

and the cutoff wavelength is related very simply to the dimensions of the guide.

We now see why it is impractical to use waveguides at the lower frequencies (longer wavelengths). We can obtain the wavelength in the guide by the same procedure used for plane waves. The fields vary as cos βz, with z (i.e., at a given instant of time), and thus the wavelength is given by $\beta\lambda_g = 2\pi$, or $\lambda_g = 2\pi/\beta = v_p/f$, where v_p is the phase velocity in the guide and f is the frequency.

We saw in the case of plane waves that it is possible to define an impedance associated with the medium, since the dimensions of E/H are impedance. We can also define an impedance for waves in a waveguide, but here there is no unique impedance associated with the guide. For a z traveling wave, the curl equations in Cartesian form (Section 10–2) give for a TE-wave:

$$\frac{\mathcal{E}_x}{\mathcal{H}_y} = -\frac{\mathcal{E}_y}{\mathcal{H}_x} = \frac{\omega\mu}{k_z},$$

and for a TM-wave

$$\frac{\mathcal{E}_x}{\mathcal{H}_y} = -\frac{\mathcal{E}_y}{\mathcal{H}_x} = \frac{k_z}{\omega\epsilon}.$$

These relations define an impedance, but the impedance is different for TE- and TM-waves. It also depends on k_z, which in turn depends on m and n. We see then that each mode has its own impedance. It is of interest to note that

$$Z^{\mathrm{TE}}Z^{\mathrm{TM}} = \frac{\mu}{\epsilon} = Z_0^2,$$

where Z_0 is the intrinsic impedance of the medium.

10–4 Power flow

We saw in Section 9–5 that the power flow density is given by the Poynting vector $\mathbf{S} = \mathbf{E} \times \mathbf{H}$. This is a product of the time-varying fields, but we have found it convenient to work with complex fields, and hence it is natural to ask whether we can define a complex counterpart of the Poynting vector. Thus far all the relations we have had between complex quantities have been linear, whereas here we have a product, which is nonlinear. The time-varying quantities are related to the complex quantities via the Re-operator, and since we know that the real part of a product is not equal to the product of the real parts we may expect some complications to arise.

Let us first consider the somewhat simpler case of voltage and current. For sinusoidal time variation we may define complex voltages and currents by the relations

$$V = \mathrm{Re}\,(\mathcal{V}e^{j\omega t}), \qquad I = \mathrm{Re}\,(\mathcal{I}e^{j\omega t}),$$

where $\mathcal{V}$ and $\mathcal{I}$ are complex and may be written $|\mathcal{V}|e^{j\alpha}$ and $|\mathcal{I}|e^{j\beta}$, and thus

$$V = \mathrm{Re}\,[|\mathcal{V}|e^{j(\omega t+\alpha)}] = |\mathcal{V}|\cos(\omega t + \alpha).$$

We see that $|\mathcal{V}|$ is the peak value of V. If the voltage and current were in phase, the power dissipated would be $|\mathcal{V}||\mathcal{I}| \cos^2 (\omega t + \alpha)$, and the time average power dissipated, $\frac{1}{2}|\mathcal{V}||\mathcal{I}|$, that is, $\overline{P} = \overline{VI} = \frac{1}{2}|\mathcal{V}||\mathcal{I}|$. If we define a complex power as $\mathcal{P} = \mathcal{V}\mathcal{I}$, then $\mathcal{P} = |\mathcal{V}||\mathcal{I}|e^{j2\alpha}$; this power is not simply related to $\overline{P}$, and we thus reject this definition. We can eliminate the $e^{j2\alpha}$-term, which is the cause of the trouble, by using $\mathcal{V}$ and $\mathcal{I}^*$ instead of $\mathcal{V}$ and $\mathcal{I}$. Hence we define the complex power $\mathcal{P}$ by the relation

$$\mathcal{P} = \mathcal{V}\mathcal{I}^*,$$

and for the case of current and voltage in phase we obtain

$$\mathcal{P} = |\mathcal{V}|e^{j(\omega t+\alpha)}|\mathcal{I}|e^{-j(\omega t+\alpha)} = |\mathcal{V}||\mathcal{I}|.$$

For current and voltage not in phase we have

$$P = \tfrac{1}{2}|\mathcal{V}||\mathcal{I}|[\cos (\alpha - \beta) + \cos (2\omega t + \alpha + \beta)],$$

or

$$\overline{P} = \tfrac{1}{2}|\mathcal{V}||\mathcal{I}| \cos (\alpha - \beta),$$

and

$$\mathcal{P} = \mathcal{V}\mathcal{I}^* = |\mathcal{V}||\mathcal{I}|e^{j(\alpha-\beta)}.$$

Hence we see that

$$\overline{P} = \tfrac{1}{2} \operatorname{Re} (\mathcal{P}),$$

and with the definition of complex power to which we were led by considering the special case of in-phase voltage and current, we have the important result that the time-average power is equal to the real part of the complex power.

We shall check whether this same result can be extended to the case of a vector cross product. We begin by considering the time-average Poynting vector

$$\overline{\mathbf{S}} = \overline{\mathbf{E} \times \mathbf{H}}.$$

By expanding the cross product in terms of its Cartesian components and using the fact that the average of the sum of a number of terms is equal to the sum of the averages, we obtain

$$\overline{\mathbf{S}} = \mathbf{u}_x(\overline{E_yH_z} - \overline{E_zH_y}) + \mathbf{u}_y(\overline{E_zH_x} - \overline{E_xH_z}) + \mathbf{u}_z(\overline{E_xH_y} - \overline{E_yH_x}).$$

The components E_y and H_z are related to the components of the complex vectors $\mathcal{E}_y$ and $\mathcal{H}_z$ in exactly the same way as V and I are related to $\mathcal{V}$ and $\mathcal{I}$. Hence by the same argument we have

$$\overline{E_yH_z} = \tfrac{1}{2} \operatorname{Re} (\mathcal{E}_y\mathcal{H}_z^*),$$

and similarly for all other terms in the expression. By using the fact that the real part of the sum of complex numbers is equal to the sum of their real parts we have

$$\bar{\mathbf{S}} = \tfrac{1}{2}\,\mathrm{Re}\,[\mathbf{u}_x(\mathcal{E}_y\mathcal{H}_z^* - \mathcal{E}_z\mathcal{H}_y^*) + \mathbf{u}_y(\mathcal{E}_z\mathcal{H}_x^* - \mathcal{E}_x\mathcal{H}_z^*) + \mathbf{u}_z(\mathcal{E}_x\mathcal{H}_y^* - \mathcal{E}_y\mathcal{H}_x^*)].$$

But the quantity in the brackets is just $\boldsymbol{\mathcal{E}} \times \boldsymbol{\mathcal{H}}^*$, and thus

$$\bar{\mathbf{S}} = \tfrac{1}{2}\,\mathrm{Re}\,(\boldsymbol{\mathcal{E}} \times \boldsymbol{\mathcal{H}}^*),$$

and hence, if we define the complex Poynting vector by the relation

$$\boldsymbol{\mathcal{S}} = \boldsymbol{\mathcal{E}} \times \boldsymbol{\mathcal{H}}^*,$$

then we have the result

$$\bar{\mathbf{S}} = \tfrac{1}{2}\,\mathrm{Re}\,(\boldsymbol{\mathcal{S}}).$$

We shall now use this result to calculate the power flowing down a waveguide. Let us first consider a guide with only the TE_{10}-mode present. To obtain the power flowing down the guide, i.e., in the z-direction, we calculate the z-component of the power flow density $\mathcal{S}_z$, and then integrate over the cross section of the guide:

$$\mathcal{S}_z = \mathcal{E}_x\mathcal{H}_y^* - \mathcal{E}_y\mathcal{H}_x^*.$$

For the TE_{10}-mode, setting $m = 1$ and $n = 0$ in the foregoing formula, we have

$$\mathcal{E}_x = \mathcal{H}_y = 0, \qquad \mathcal{E}_y = \mathcal{E}_0 \sin\frac{\pi}{a}x\; e^{-jk_z z}, \qquad \mathcal{H}_x = -\frac{k_z}{\omega\mu}\,\mathcal{E}_y;$$

therefore

$$\mathcal{S}_z = \frac{k_z}{\omega\mu}\,\mathcal{E}_y\mathcal{E}_y^* = \frac{k_z}{\omega\mu}\,|\mathcal{E}_y|^2 = \frac{k_z}{\omega\mu}\,\mathcal{E}_0^2\sin^2\frac{\pi}{a}x.$$

The time-average power flowing across an element of area of the guide $dx\,dy$ is given by $\bar{S}_z\,dx\,dy$, and that flowing down the whole guide by

$$\int_0^a\int_0^b \bar{S}_z\,dx\,dy.$$

We see that

$$\bar{S}_z = \tfrac{1}{2}\,\mathrm{Re}\,(\mathcal{S}_z) = \frac{1}{2}\frac{k_z}{\omega\mu}\,\mathcal{E}_0^2\sin^2\frac{\pi}{a}x.$$

The integration required is therefore

$$\int_0^a\int_0^b \sin^2\frac{\pi}{a}x\,dx\,dy = \tfrac{1}{2}ab,$$

and hence the power flowing down the guide is $\frac{1}{4}ab(k_z/\omega\mu)\,\mathcal{E}_0^2$.

Next let us suppose that there are two TE-modes propagating down the guide simultaneously. Since the medium in the guide is assumed to be linear, the principle of superposition applies, and the resulting **E**- and **H**-fields are simply the sum of the component fields. The complex power flow in the z-direction is given by

$$\mathcal{S}_z = \mathcal{E}_x \mathcal{H}_y^* - \mathcal{E}_y \mathcal{H}_x^*.$$

Since

$$\mathcal{E}_x = A \cos \frac{m\pi}{a} x \sin \frac{n\pi}{b} y \, e^{-jk_z z} + B \cos \frac{k\pi}{a} x \sin \frac{l\pi}{b} y \, e^{-jk'_z z},$$

where A and B contain the magnitudes and phases of the $\text{TE}_{m,n}$- and $\text{TE}_{k,l}$-modes, and

$$\mathcal{H}_y = A \frac{k_z}{\omega\mu} \cos \frac{m\pi}{a} x \sin \frac{n\pi}{b} y \, e^{-jk_z z} + B \frac{k'_z}{\omega\mu} \cos \frac{k\pi}{a} x \sin \frac{l\pi}{b} y \, e^{-jk'_z z},$$

then for $\mathcal{E}_x \mathcal{H}_y^*$, the first of the four terms in $\mathcal{S}_z$, we have

$$A \cos \frac{m\pi}{a} x \sin \frac{n\pi}{b} y \, e^{-jk_z z} A^* \frac{k_z}{\omega\mu} \cos \frac{m\pi}{a} x \sin \frac{n\pi}{b} y \, e^{jk_z z}.$$

The power flow down the guide is given by $\int_0^a \int_0^b \mathcal{S}_z \, dx \, dy$, and integrating this term gives

$$|A|^2 \frac{k_z}{\omega\mu} \int_0^a \int_0^b \cos^2 \frac{m\pi}{a} x \sin^2 \frac{n\pi}{b} y \, dx \, dy = \tfrac{1}{4} |A|^2 \frac{k_z}{\omega\mu} ab,$$

which would be the first term $\mathcal{E}_x \mathcal{H}_y^*$ in the expression for the power carried by the $\text{TE}_{m,n}$-mode, if it existed alone in the guide. Similarly we have the term

$$\tfrac{1}{4} |B|^2 \frac{k'_z}{\omega\mu} ab,$$

which would be the first term of the power carried by the $\text{TE}_{k,l}$-mode alone. We also have the term

$$AB^* \frac{k_z}{\omega\mu} \int_0^a \int_0^b \cos \frac{m\pi}{a} x \cos \frac{k\pi}{a} x \sin \frac{n\pi}{b} y \sin \frac{l\pi}{b} y \, e^{-j(k_z - k'_z) z} \, dx \, dy,$$

which is zero. The fourth term in the expression also gives a zero contribution to the integral. A similar treatment can be applied to the second term of the expression, $\mathcal{E}_y \mathcal{H}_x^*$, and we therefore obtain the result that the total power flow down the guide is the sum of the powers carried by the two modes. This result can clearly be extended to any number of modes.

We shall now calculate the power flow down the guide when the wave is evanescent. In this case the z-variation is given by $e^{-\alpha z}$. For the $\text{TE}_{m,n}$-mode, putting $k_z = -j\alpha$, we have

$$\mathcal{E}_x = A \cos \frac{m\pi}{a} x \sin \frac{n\pi}{b} y \, e^{-\alpha z},$$

$$\mathcal{H}_y^* = A^* \frac{j\alpha}{\omega\mu} \cos \frac{m\pi}{a} x \sin \frac{n\pi}{b} y \, e^{-\alpha z}.$$

Hence the product $\mathcal{E}_x\mathcal{H}_y^*$ is a pure imaginary. Similarly it may be shown that $\mathcal{E}_y\mathcal{H}_x^*$ is a pure imaginary, and so Re $(\mathcal{S}_z) = 0$, and we see that the time-average power flow down the guide is zero. Thus the evanescent mode in a waveguide below cutoff transfers no power.

Turning now to the time-varying quantities and taking the TE_{10}-mode, we have

$$E_y = A \sin \frac{\pi}{a} x \, e^{-\alpha z} \cos \omega t, \qquad H_x = A \frac{\alpha}{\omega\mu} \sin \frac{\pi}{a} x \, e^{-\alpha z} \sin \omega t,$$

where we have chosen the origin of t so that the phase angle is zero, and thus A is real. Hence

$$S_z = -E_y H_x = -\tfrac{1}{2} A^2 \frac{\alpha}{\omega\mu} e^{-2\alpha z} \sin^2 \frac{\pi}{a} x \sin 2\omega t,$$

and integrating over the cross section of the guide gives the power flow

$$-\tfrac{1}{4} ab A^2 \frac{\alpha}{\omega\mu} e^{-2\alpha z} \sin 2\omega t.$$

There is, therefore, an oscillation of energy in the z-direction at a frequency equal to twice the wave frequency.

The orthogonality property of the modes across the cross section of the guide is very important. It leads, as we have just seen, to the result that the power flow down the guide is the sum of the powers carried by the individual modes. It also enables us to synthesize any arbitrary field specified across a cross section of the guide in terms of these modes. The amplitudes of the modes can be calculated in a way similar to that used to determine the amplitudes of the harmonics when we analyze a complex function of time. It is more complicated in this case, however, since we must use a double Fourier series instead of a single one, as the function we are analyzing is dependent on two variables, x and y. The TE- and TM-modes are not the most suitable sets for these analyses. We saw that of the six components of the field we could choose E_z or H_z to be zero and still have a nonzero field in the guide. Actually *any* one component can be chosen to be zero. Modes in which one of the transverse components is zero are known as *hybrid modes* and are the most useful modes for synthesizing specified fields. A complete treatment of modal analysis, however, is beyond the scope of this text.

10–5 Equivalent circuit

We saw in the previous section that each mode in a waveguide has a power flow associated with it above cutoff. We also know from Section 10–2 that we can associate an impedance with each mode. Thus we can define a voltage and a current associated with each mode from the relations

$$P = VI, \qquad Z = \frac{V}{I} \qquad \text{or} \qquad V = \sqrt{PZ}, \qquad I = \sqrt{P/Z}.$$

If we can define a voltage and a current, then we can work in terms of an equivalent circuit, and all the ideas of circuit theory will apply. Note that we have not produced an equivalent circuit for the waveguide, but only for a mode in the waveguide. This idea is of greatest use when the waveguide is designed so that only the dominant mode propagates. In this case we have only the one mode to deal with for the most part, and thus the simple equivalent circuit applies. This is not quite the whole story, however. A more detailed analysis shows that discontinuities in the waveguide such as occur at waveguide junctions excite the higher-order modes. These are evanescent and have appreciable magnitude only near the junctions. If these evanescent modes do not interact with any other part of the system, they can be neglected. For example, let us suppose that we have a waveguide connecting two cavities. At the junction between one cavity and the guide, the evanescent fields are excited, but their magnitude decreases rapidly as we move along the waveguide toward the other cavity. If the interconnecting guide is long enough so that the magnitude of these fields is negligible by the time the other junction is reached, there is no interaction between the cavities due to the evanescent fields. The only interaction is due to the propagating dominant mode, and hence the equivalent circuit will be an accurate representation of the system; this circuit is widely used in the analysis of complex microwave systems and forms the basis of the concept of microwave circuits.

10–6 The cavity resonator

Let us investigate the types of field which can exist in a rectangular box, as shown in Fig. 10–3. We shall still have the solution of the field equations $X(x)Y(y)Z(z)$, where X, Y, and Z satisfy the simple harmonic equation we obtained in Section 10–2. The boundary conditions, however, will now be that the tangential component of $\mathbf{E}$ must be zero over all six faces of the box, i.e., at $x = 0, a$; $y = 0, b$; $z = 0, c$. This problem is most simply solved by returning to a consideration of the waveguide. We have already derived the traveling waves which can exist in the rectangular guide with sides a and b. For example, for the $\mathcal{E}_x$-field of the TE_{mn}-mode, we had

$$\mathcal{E}_x = A \cos \frac{m\pi}{a} x \sin \frac{n\pi}{b} y \, e^{-jk_z z}.$$

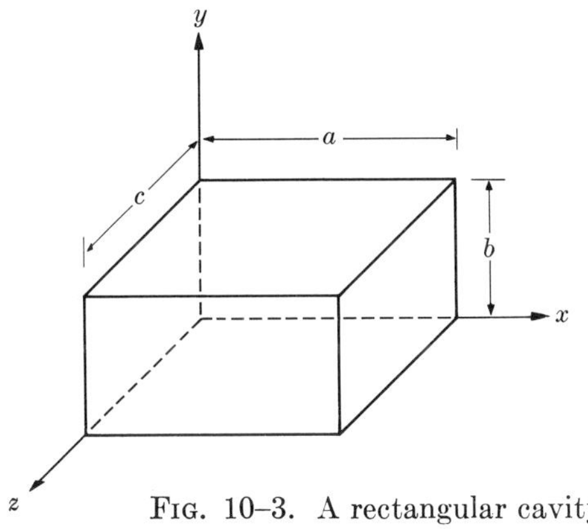

FIG. 10-3. A rectangular cavity.

We can immediately see that the above expression cannot be made to fit the boundary conditions for the problem; it represents a field traveling in the $+z$-direction. If we add to it a wave traveling in the $-z$-direction, we produce a standing wave in the guide. To fit the boundary conditions for z we need an amplitude $-A$ for the negatively traveling field. This gives a resultant field of

$$\mathcal{E}_x = B \cos \frac{m\pi}{a} x \sin \frac{n\pi}{b} y \sin k_z z,$$

which satisfies the condition that $\mathcal{E}_x$ (which is one of the tangential components of $\boldsymbol{\mathcal{E}}$ over the faces perpendicular to z) is zero at $z = 0$. We must also have $\mathcal{E}_x = 0$ at $z = c$, and this requires $k_z = p\pi/c$. Hence we finally have

$$\mathcal{E}_x = B \cos \frac{m\pi}{a} x \sin \frac{n\pi}{b} y \sin \frac{p\pi}{c} z.$$

A detailed investigation shows that this choice of solution will satisfy all the boundary conditions, that is, $\mathcal{E}_x = 0$ at $y = b$ and $z = c$, $\mathcal{E}_y = 0$ at $x = a$ and $z = c$, and $\mathcal{E}_z = 0$ at $x = a$ and $y = b$. The parameter p can have any integral value except zero. This set of fields in the cavity is designated the $\text{TE}_{m,n,p}$-set, and we can have $m = 0, 1, 2, \ldots$, $n = 0, 1, 2, \ldots$, and $p = 1, 2, 3, \ldots$, with the exception of the case $m = n = 0$. From the separation equation we have $k^2 = k_x^2 + k_y^2 + k_z^2$, and for a lossless medium in the cavity, $\omega^2\mu\epsilon = (m\pi/a)^2 + (n\pi/b)^2 + (p\pi/c)^2$. Thus the frequencies of the oscillations which can exist in the cavity are restricted to the values

$$\omega = (1/\sqrt{\mu\epsilon})\sqrt{(m\pi/a)^2 + (n\pi/b)^2 + (p\pi/c)^2},$$

which are known as the resonant frequencies of the cavity. Since we have assumed no losses, these resonances are infinitely sharp. In practice, of course, there will always be some loss in the cavity, the most important cause being the fact that the walls of the guide are not perfect conductors. The cavity is the microwave counterpart of the tuned circuit. Note, however, that the cavity has an infinite number of resonant frequencies, whereas the tuned circuit has only one. The resonant frequency $f_{m,n,p}$ is given by

$$f_{m,n,p} = \sqrt{(m/a)^2 + (n/b)^2 + (p/c)^2}/2\sqrt{\mu\epsilon}.$$

We let $a = 2$, $b = 1$, and $c = 2$. The resonant frequencies for the first few TE-modes are:

mnp	101	011	201	102	111	012
$2\sqrt{\mu\epsilon}\, f_{m,n,p}$	$\sqrt{\frac{1}{2}}$	$\sqrt{\frac{5}{4}}$	$\sqrt{\frac{5}{4}}$	$\sqrt{\frac{5}{4}}$	$\sqrt{\frac{3}{2}}$	$\sqrt{2}$

The mode with the lowest resonant frequency is the TE_{101}-mode. The next lowest is $\sqrt{5/2}$ or 1.58 times this, and three TE-modes resonate at this frequency. An exhaustive investigation of the possibilities shows that this is the greatest separation which can be achieved between the dominant mode resonance and the next lowest-order mode resonance. The foregoing modes of the cavity were based on the TE-solutions in the guide. Another set can be obtained by superimposing a $+z$ and a $-z$ traveling TM-wave. We have noted that the lowest-order TM-mode is $m = n = 1$. If all the field components are written out, then it may be seen that $p = 0$ gives a nonzero field, and hence the modes possible are $TM_{m,n,p}$ with $m = 1, 2, 3, \ldots$, $n = 1, 2, 3, \ldots$, and $p = 0, 1, 2, \ldots$ The lowest-order cutoff is given by $m = n = 1$, $p = 0$, and for the cavity discussed earlier this yields $2\sqrt{\mu\epsilon}\, f_{110} = \sqrt{5/4}$, which is removed from the dominant mode resonance by the factor $\sqrt{5/2}$. Hence the TM-modes do not produce resonances which are closer to the dominant mode resonance than those produced by the higher-order TE-modes.

We shall now investigate the energy in the TE_{101}-fields. For this mode,

$$E_x = E_z = H_y = 0.$$

The remaining three components are given by

$$E_y = E_0 \sin\frac{\pi}{a}x \sin\frac{\pi}{c}z \cos\omega t,$$

$$H_x = E_0 \frac{1}{\omega\mu}\frac{\pi}{c} \sin\frac{\pi}{a}x \cos\frac{\pi}{c}z \sin\omega t,$$

$$H_z = -E_0 \frac{1}{\omega\mu}\frac{\pi}{a} \cos\frac{\pi}{a}x \sin\frac{\pi}{c}z \sin\omega t.$$

We may note immediately that the **E**- and **H**-fields are in phase quadrature just as the voltage and current are in a tuned circuit at resonance. The electric energy density in the cavity is $\frac{1}{2}\epsilon E^2$, and hence the total energy is

$$\int_0^a \int_0^b \int_0^c \tfrac{1}{2}\epsilon E^2 \, dx \, dy \, dz = \tfrac{1}{2}\epsilon E_0^2 \cos^2 \omega t \int_0^a \int_0^b \int_0^c \sin^2 \frac{\pi}{a} x \sin^2 \frac{\pi}{c} z \, dx \, dy \, dz$$

$$= \tfrac{1}{8} abc\epsilon E_0^2 \cos^2 \omega t.$$

The total magnetic energy is given by

$$\int_0^a \int_0^b \int_0^c \tfrac{1}{2}\mu(H_x^2 + H_z^2) \, dx \, dy \, dz$$

$$= \tfrac{1}{2}\mu E_0^2 \sin^2 \omega t \frac{\pi^2}{\omega^2 \mu^2} \int_0^a \int_0^b \int_0^c \left(\frac{1}{c^2} \sin^2 \frac{\pi}{a} x \cos^2 \frac{\pi}{c} z + \frac{1}{a^2} \cos^2 \frac{\pi}{a} x \sin^2 \frac{\pi}{c} z\right) dx \, dy \, dz$$

$$= \tfrac{1}{2}\mu E_0^2 \sin^2 \omega t \frac{\pi^2}{\omega^2 \mu^2} \tfrac{1}{4} abc \left(\frac{1}{c^2} + \frac{1}{a^2}\right),$$

but for the TE_{101}-mode, we have

$$\omega = \pi\sqrt{1/a^2 + 1/c^2}/\sqrt{\mu\epsilon}.$$

Hence the magnetic energy is given by

$$\tfrac{1}{8} abc\epsilon E_0^2 \sin^2 \omega t.$$

We note that the sum of the electric and magnetic energies is $\frac{1}{8}abc\epsilon E_0^2$, which is constant, as it must be by the law of conservation of energy, since the system is supposed to be lossless. When the energy is a maximum in the electric field, it is zero in the magnetic field and vice versa. This is a general characteristic of resonant systems. There are two methods by which energy can be stored, and the energy oscillates back and forth between them. For example, in a tuned circuit the energy oscillates between the capacitor and the inductor.

10–7 Microwave systems

We have so far considered the waveguide and the cavity resonator. These are the microwave counterparts of the connection wires and the tuned circuits and thus comprise the linear passive bilateral parts of the system. We have not yet considered amplifiers and oscillators at microwaves. Accordingly we have as yet no clear picture of what constitutes a source at microwaves. We shall not go into this matter here, but we might note that the commonly used oscillators, such as the magnetron and the klystron, consist basically of a cavity in which the oscillations are maintained by various feedback devices. We obviously cannot make use of these fields (or, for that matter, maintain them) in a completely closed cavity. We must either put in a probe or cut an aperture in the cavity so that we can couple to these fields. Such a system becomes very difficult to handle exactly by mathematical analysis of the fields. We might consider, for example, cutting a rectangular hole in the cavity and fixing a rectangular waveguide on to it. We must note that the presence of the guide will affect the fields in the cavity. We can regard this as a loading effect on the source. The magnitude of the effect of the waveguide on the cavity will depend on how tightly the fields in the cavity and the waveguide are coupled. This type of problem is analyzed by making use of the equivalent circuits. For details, the reader is referred to texts on microwave circuits.

Problems

10-1 Starting from the wave equation, show that the complete solution for the TE-wave of Section 10-2 is

$$\begin{aligned}
\mathcal{E}_x &= \frac{j\omega\mu}{k^2 - k_z^2}\frac{n\pi}{b}\cos\frac{m\pi}{a}x\,\sin\frac{n\pi}{b}y\;e^{-jk_z z},\\
\mathcal{E}_y &= -\frac{j\omega\mu}{k^2 - k_z^2}\frac{m\pi}{a}\sin\frac{m\pi}{a}x\,\cos\frac{n\pi}{b}y\;e^{-jk_z z},\\
\mathcal{E}_z &= 0,\\
\mathcal{H}_x &= \frac{jk_z}{k^2 - k_z^2}\frac{m\pi}{a}\sin\frac{m\pi}{a}x\,\cos\frac{n\pi}{b}y\;e^{-jk_z z},\\
\mathcal{H}_y &= \frac{jk_z}{k^2 - k_z^2}\frac{n\pi}{b}\cos\frac{m\pi}{a}x\,\sin\frac{n\pi}{b}y\;e^{-jk_z z},\\
\mathcal{H}_z &= \cos\frac{m\pi}{a}x\,\cos\frac{n\pi}{b}y\;e^{-jk_z z}.
\end{aligned}$$

10-2 Verify that the boundary condition for **E** (i.e., tangential electric field is zero) is satisfied over all four walls of the guide.

10-3 Derive the complete set of relations for the TM-wave starting from the wave equation.

10-4 Investigate thoroughly the boundary conditions for the TM-wave. Calculate the currents in the walls of the guide.

10-5 Show that in the limit as $b \to \infty$ so that the guide becomes a parallel-plate transmission line, a TEM-wave can exist with a finite amplitude $\mathcal{E}_x$.

10-6 Determine the phase velocity and the group velocity of the wave at the cutoff frequency.

10-7 Calculate the cutoff frequencies for the five lowest-order TE-waves and the five lowest-order TM-waves for a waveguide with $a = 2b$ (Section 10-3). Express your results in terms of the cutoff frequency of the lowest-order mode.

10–8 A rectangular waveguide measures 8 cm $\times$ 4 cm. Calculate the lowest cutoff frequency. For a frequency 1.5 times this minimum, calculate the phase velocity, the group velocity, and the wavelength in the guide. What dimensions would be required for a waveguide with sides in the ratio of 2 : 1 for a wavelength of 1 mm, if only the dominant mode propagates?

10–9 A waveguide contains two modes, one of which is propagating and one is evanescent. Calculate the Poynting vector (time-dependent form) for this case and obtain the power flow down the guide.

10–10 Make a careful check of the boundary conditions for the cavity resonator discussed in Section 10–6.

10–11 Make a table of resonant frequencies for the TM-modes of the cavity similar to that given in Section 10–6.

CHAPTER 11

The Motion of Charged Particles in Fields

11–1 Introduction

In previous chapters we considered the forces acting on electrostatic charges and electric currents in electric and magnetic fields. Here we shall extend this treatment by considering charged particles. The particle is considered to have not only an electric charge, but also a mass. In addition to the laws of electromagnetism, we introduce the laws of dynamics, which enable us to calculate the trajectories of the particles.

Let us suppose that we have a particle of mass m carrying a charge q. If this particle is in a uniform electrostatic field, then we know from Chapter 3 that the force acting on it is given by $q\mathbf{E}$. The particle will therefore have an acceleration $(q/m)E$, that is, $dv/dt = (q/m)E$. Since $v = dl/dt$, we can write

$$v \frac{dv}{dt} = \frac{1}{2} \frac{d}{dt} (v^2) = \frac{qE}{m} \frac{dl}{dt},$$

and, if the particle begins at rest, integration yields

$$v^2 = \frac{2q}{m} El.$$

Note that since the only force acting on the particle is due to the field $\mathbf{E}$, the particle will move in the direction of the field, that is, $\mathbf{E}$ and $d\mathbf{l}$ are in the same direction. But for a uniform field, $El = V_0 - V$, where V_0 is the potential at the initial point and V that at the final point. Since the zero of potential is arbitrary, we let $V_0 = 0$, and if we consider electrons so that the charge is negative, $q = -e$, we have

$$v = \sqrt{2eV/m},$$

which gives the velocity acquired by an electron in accelerating from zero velocity to a potential V. The same result can, of course, be obtained from energy considerations. If an electron falls through a potential V, then it loses potential energy eV, since the work done in moving a charge e through a potential difference V is eV. This must be transferred into kinetic energy, and hence $\frac{1}{2}mv^2 = eV$ or $v = \sqrt{2eV/m}$. When the electron, which has a charge of 1.6×10^{-19} coul, falls through a potential difference of 1 v, it acquires an energy of 1.6×10^{-19} *joule* (j). This is a convenient unit of energy in atomic physics and is referred to as 1 electron volt (1 eV).

We know that a moving charge constitutes an electric current, and consequently if a charged particle is in motion in a magnetic field, there will be a force on it. We saw in Chapter 4 that the force on a current element of length $d\mathbf{l}$ and current I in a magnetic field can be written $\mathbf{F} = I\,d\mathbf{l} \times \mathbf{B}$. Let us consider a uniform charge density q_v moving with a uniform velocity $\mathbf{v}$. The charge density crossing an area dA perpendicular to the direction of motion is $q_v v\,dA$ per unit time. Hence for the force, we have

$$\mathbf{F} = q_v v\,dA\,d\mathbf{l} \times \mathbf{B}.$$

Note that $\mathbf{v}$ and $d\mathbf{l}$ are in the same direction, and hence this expression can be written

$$\mathbf{F} = q_v\,dA\,dl\,\mathbf{v} \times \mathbf{B}.$$

But since dA is perpendicular to $d\mathbf{l}$, $dA\,dl$ is just the volume of the element, and since q_v is the charge density per unit volume, we have

$$\mathbf{F} = q\mathbf{v} \times \mathbf{B},$$

where q is the charge on the element. The complete expression for the force on a charge in the presence of electric and magnetic fields then becomes

$$\mathbf{F} = q(\mathbf{E} + \mathbf{v} \times \mathbf{B}).$$

Continuing with the case of the magnetic field alone, we see that the force is always perpendicular to the magnetic field $\mathbf{B}$ and to the direction of motion of the particle $\mathbf{v}$, since these appear in the expression as a cross product. We know from mechanics that if the force is always directed perpendicular to the motion, then the particle will describe a circle. The centrifugal force on a particle moving with speed v in a circle of radius r is mv^2/r. The velocity of the particle can be resolved into two components, a longitudinal component $v_{||}$ parallel to the field $\mathbf{B}$, and a transverse component $v_\perp$ perpendicular to it. The longitudinal component $v_{||}$ is unaffected by the presence of the field, since the force is perpendicular to $\mathbf{B}$ and $\mathbf{v} \times \mathbf{B}$ is zero. Equating the centrifugal force to the force on the particle, we obtain

$$\frac{mv_\perp^2}{r} = qv_\perp B,$$

and the radius of the circle is

$$r = \frac{mv_\perp}{qB}.$$

The angular velocity of the particle is given by $v_\perp/r$, and so

$$\omega = \frac{qB}{m}.$$

This frequency is variously known as the cyclotron, gyro, or Larmor frequency. The resulting motion of a charged particle in a magnetic field is composed of a steady motion along the field and a circular motion around the field. The particle therefore moves along a helix whose axis is parallel to the field.

11–2 The cathode-ray tube

One of the most direct applications of the results of the last section is the cathode-ray tube. Here we have a beam of electrons which impinges on a sensitive screen and produces a spot of light. In an electrostatic tube, the position of the spot is controlled by two pairs of deflector plates, one for the horizontal position and one for the vertical. The spot may be deflected by putting voltage across the plates, which produces a field between the plates, and as the electrons pass through this region they experience forces as described above. Let us consider the situation shown in Fig. 11–1. We will suppose that the field due to the voltage V_y across the deflector plates is confined to the region between the plates; i.e., we neglect edge effects. Hence the electron is acted on by a force eE during its passage through the plates. Thus the equation of motion is given by

$$m\frac{dv_y}{dt} = eE, \qquad \text{and} \qquad v_y = \frac{eE}{m}t,$$

since $v_y = 0$ before the electron enters the field, and we can assume that $t = 0$ at the instant the electron enters the field. This force acts for a time equal to the time it takes the electron to travel the distance s. Since the field is perpendicular to the axis, the axial velocity of the electron is constant, and this time is therefore given by s/v. If the voltage between the deflector plates and the cathode is V, then

$$\tfrac{1}{2}mv^2 = eV.$$

Finally, assuming $s \ll l$, we have $y/l = v_y/v$. Hence

$$y = l\frac{v_y}{v} = \frac{leEs}{mv^2} = \frac{lEs}{2V}.$$

If the deflector plates are a distance d apart, then $E = V_y/d$, and

$$y = \frac{ls}{2d}\frac{V_y}{V}.$$

The *deflection sensitivity* is defined as the deflection per unit potential difference between the plates and is given by $ls/2Vd$. Increasing s or decreasing d will increase the sensitivity; however, this cannot be carried too far or the electrons will strike the deflector plates. Increasing l, that is, making the tube longer, will increase the sensitivity, and the sensitivity decreases with increasing accelerator voltage V.

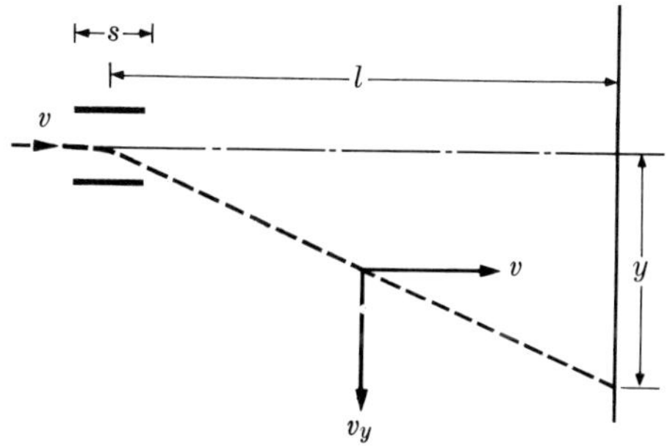

FIG. 11-1. The electrostatic deflection system.

The calculation above has been made on the assumption of a steady deflection voltage. In practice, we often want to use the cathode-ray tube to examine time-varying voltages. If we let $V_y = V_0 \cos \omega t$, then

$$m \frac{dv_y}{dt} = \frac{e}{d} V_0 \cos \omega t.$$

The axial velocity is still unaffected by this field, and thus we have $t = s/v$. Hence

$$v_y = \frac{eV_0}{md} \int_0^{s/v} \cos \omega t \, dt = \frac{eV_0}{m\omega d} \sin \frac{\omega s}{v} \cdot$$

Note that as $\omega \rightarrow 0$ this expression reduces to the same result as before. We see that the deflection sensitivity which is proportional to v_y/V_0 varies like $(\sin x)/x$, and thus will become zero when $\omega s/v = \pi$ or when $f = \omega/2\pi = v/2s$. We can obtain the velocity of the electrons from the formula derived above:

$$v = \sqrt{2eV/m}.$$

The charge on the electron is 1.6×10^{-19} coul, and its mass is 0.91×10^{-30} kg, so that $v = 5.93 \times 10^5 \sqrt{V}$ m/sec. If the accelerating voltage is 2500 v, then $v \doteq 3 \times 10^7$ m/sec. If the length of the deflector plate is 1.5 cm, $f = 10^9$ cycles/sec or 1000 Mc/sec. We see that at the cutoff frequency the time for the applied voltage to go through half a cycle is equal to the time taken for the electron to travel the length of the deflector plate. In practice, these transit-time effects may be entirely neglected for frequencies less than 100 Mc/sec.

We can also use a magnetic field to deflect the beam. To produce a vertical deflection (assuming the electron beam to be horizontal) we would need a horizontal magnetic field perpendicular to the beam. This could be produced by a pair of coils, as shown in Fig. 11-2. The deflection may be obtained in terms of the field as follows. We will assume that the magnetic field is perpendicular to the page (Fig. 11-3), that it is confined to the region of the deflector

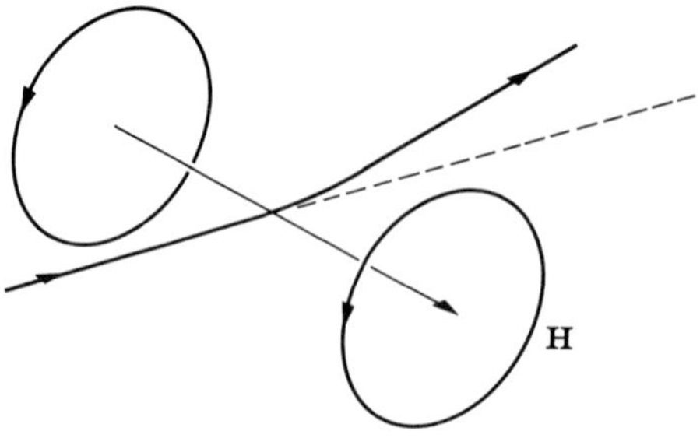

FIG. 11-2. The electromagnetic deflection system.

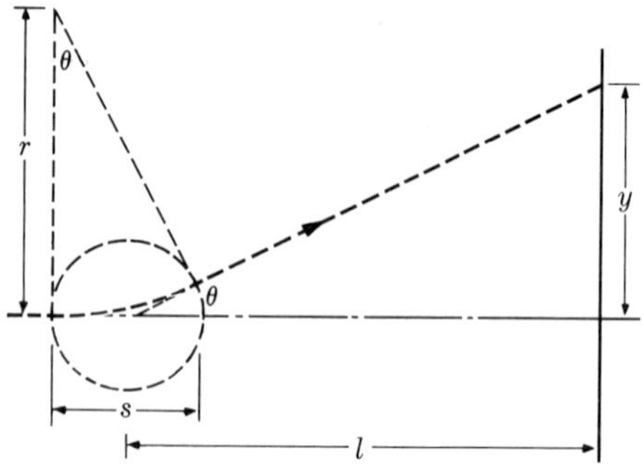

FIG. 11-3. The calculation of the deflection.

assembly (no edge effects), and that it is uniform in this region. We will further suppose that $l \gg s$ and that θ is a small angle. While the electron is in the field, its trajectory is a circle whose radius we have determined to be mv/eB. The angle θ is therefore given by $s/r = eBs/mv$. The deflection may be obtained from the relation $y = \theta l$. Therefore

$$y = \frac{eBsl}{mv},$$

and since

$$v = \sqrt{2eV/m},$$

hence

$$y = (lBs/\sqrt{V})\sqrt{e/2m}.$$

This expression has some similarities to that for electrostatic deflection. We see that the deflection is increased by increasing s, the size of the deflector assembly, and by increasing l, the length of the tube. It is decreased by increasing the accelerating voltage V, but in this case, not so fast, since the deflection varies with the inverse of the square root of V. In the case of magnetic deflection, the deflection depends on the ratio of the charge to the mass of the particle, whereas this factor did not enter into the expression for electrostatic deflection. The deflection is proportional to B, which will in turn be proportional to the current in the coil producing it. Thus magnetic deflection gives a deflection proportional to current. Electrostatic deflection, on the other hand, gives a deflection proportional to voltage.

We have referred above to the electron beam in the cathode-ray tube. The source of the electrons is thermionic emission from a cathode as in the vacuum tube, and these electrons can be accelerated by a positive voltage. However, they will not automatically form a beam; in fact, owing to their mutual repulsion, the electrons tend to diverge. To form a beam we must use some sort of focusing. Since we have already seen that charged particles can be deflected by electric

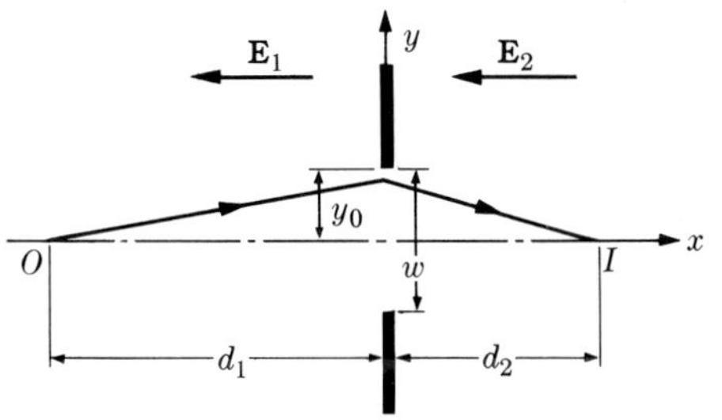

FIG. 11–4. The electrostatic lens.

and magnetic fields, we naturally look to the use of these fields to provide focusing. Let us first consider an infinite metal plate with a narrow slit in it (see Fig. 11–4). We will suppose the slit to be at $x = 0$, to have a width w in the y-direction, and to be infinite in the z-direction. We also have a uniform electric field E_1 to the left of the plate and E_2 to the right of it. We suppose that there is a source of electrons at O. These will be accelerated by the field E_1 toward the slit. Owing to the random nature of the emission, these electrons will have a velocity in the y-direction. Let us next consider an electron which has a trajectory making a small angle with the x-axis, as shown in the figure. To calculate what happens to this electron as it passes through the slit, we begin with the equation for the potential, i.e., Laplace's equation. (We suppose that the field due to the electrons themselves is small compared with E_1 and E_2.) Since the plate and the slit are infinite in the z-direction, we assume that the solution will be the same for all values of z, that is, $\partial/\partial z = 0$. Hence Laplace's equation becomes

$$\frac{\partial^2 V}{\partial x^2} + \frac{\partial^2 V}{\partial y^2} = 0.$$

We will make the further assumption that there is no component of the field in the y-direction except in the immediate vicinity of the slit, so that any change in the y-component of velocity will take place near the origin of x and y (assuming that the slit is narrow). The force causing this change of motion is $e(\partial V/\partial y)$. (Since the charge on the electron is negative and the field is $-\partial V/\partial y$, this expression gives the force in the $+y$-direction.) Because we are concerned with the region near the origin, we can expand $\partial V/\partial y$ as a Taylor series:

$$\frac{\partial V}{\partial y} = \left(\frac{\partial V}{\partial y}\right)_0 + y\left(\frac{\partial^2 V}{\partial y^2}\right)_0 + \cdots$$

Since $e(\partial V/\partial y)_0$ is the value of the force at $y = 0$, that is, on the axis of the system, then from the symmetry of the system this is zero. Hence, neglecting higher-order terms, we find that the force is given by

$$e\frac{\partial V}{\partial y} = ey\left(\frac{\partial^2 V}{\partial y^2}\right)_0.$$

In the region close to the origin we can write, to a first approximation,

$$\left(\frac{\partial^2 V}{\partial y^2}\right)_0 = \frac{\partial^2 V}{\partial y^2},$$

and hence

$$e\,\frac{\partial V}{\partial y} = -ey\,\frac{\partial^2 V}{\partial x^2}.$$

By Newton's law this is equal to mass times acceleration, and thus

$$m\,\frac{d^2 y}{dt^2} = -ey\,\frac{\partial^2 V}{\partial x^2}.$$

But we have

$$\frac{dy}{dt} = \frac{dy}{dx}\frac{dx}{dt}.$$

We see that dx/dt is the velocity in the x-direction, and if the angle of the trajectory with the x-axis is small, this is approximately equal to v, the velocity of the electrons. Hence

$$\frac{dy}{dt} = v\,\frac{dy}{dx}.$$

Assuming v is approximately constant, we obtain

$$\frac{d^2 y}{dt^2} = v^2\,\frac{d^2 y}{dx^2}, \qquad \frac{d^2 y}{dx^2} = -\frac{ey}{mv^2}\frac{\partial^2 V}{\partial x^2}.$$

If we suppose that, for a small region near the origin, y can be considered independent of x, then on integrating once and putting $mv^2 = 2eV$ we have

$$\left(\frac{dy}{dx}\right)_1 - \left(\frac{dy}{dx}\right)_2 = \left(-\frac{y_0}{2V}\frac{\partial V}{\partial x}\right)_1 - \left(-\frac{y_0}{2V}\frac{\partial V}{\partial x}\right)_2.$$

But $(dy/dx)_1 = y_0/d_1$ and $(dy/dx)_2 = -y_0/d_2$, and with E_1 and E_2 in the directions shown in Fig. 11-6, $(\partial V/\partial x)_1 = E_1$ and $(\partial V/\partial x)_2 = E_2$. Substituting we obtain

$$\frac{1}{d_1} + \frac{1}{d_2} = \frac{E_2 - E_1}{2V}.$$

Since the result is independent of y_0, all electrons leaving O within a small angle of the axis will pass through I. Thus the slit has a focusing action. In fact, the above relation is recognized to be the same as the formula for a thin lens with a focal length $2V/(E_2 - E_1)$. The approximations involved are similar. For the lens, "thin" implies that all the refraction of the light takes place in a small region. Here we have supposed that the force on the electron is confined to the region near the slit. In optics we consider only rays near the

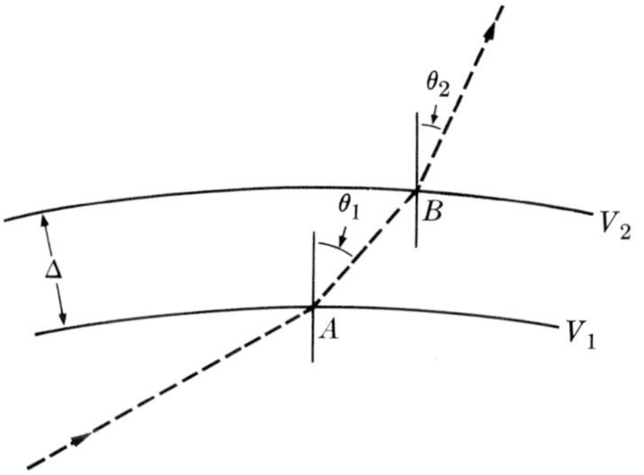

FIG. 11–5. The trajectory of an electron in an electrostatic field.

axis of the lens in order to get a point focus and not a caustic curve. Here we consider electron trajectories which make only small angles with the axis. Note that if $E_2 > E_1$, the focal length is positive, and the lens is a converging lens, whereas if $E_1 > E_2$, the focal length is negative, and the lens is diverging.

The analogy between the electron behavior and light has led to the subject of "electron optics." The simple lens in optics is, in general, inadequate, and lenses for cameras or telescopes are much more complicated and are designed by ray-tracing techniques and empirical methods. Similarly, the electron lenses in practical use in cathode-ray tubes are more complex structures, and empirical methods are used in their design. The analogy between electron beams and light beams can be extended further; we shall next derive the counterpart of Snell's law.

Let us consider two equipotential surfaces V_1 and V_2 separated by a small distance Δ, as shown in Fig. 11–5. An electron trajectory is shown crossing these surfaces at A and B. The velocity vectors v_1 and v_2 at A and B will be tangential to the trajectory, and the vectors make angles θ_1, θ_2 with the normals to the equipotential surfaces at A and B. We suppose Δ to be small enough for these two normals to be considered parallel. The force on the electron is in the direction of the electric field which is normal to the equipotentials. Therefore the component of velocity parallel to the equipotentials is constant, and hence

$$v_1 \sin \theta_1 = v_2 \sin \theta_2.$$

But for the velocity of the electron we have

$$v = \sqrt{2eV/m},$$

and thus

$$\sqrt{V_1} \sin \theta_1 = \sqrt{V_2} \sin \theta_2,$$

which is analogous to Snell's law, with $\sqrt{V}$ taking the place of the refractive index. To trace a trajectory through an electron lens it is first necessary to obtain the equipotential surfaces. For the complicated type of structure needed for electron lenses, this cannot be done analytically. Various methods have

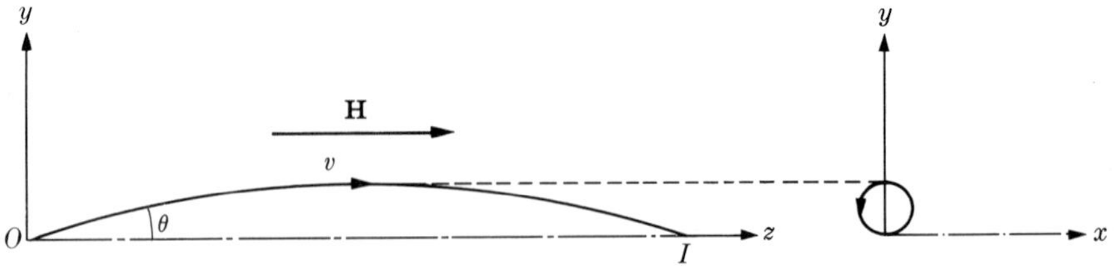

FIG. 11-6. Magnetic focusing.

been developed for obtaining the equipotentials.* This way of looking at the action of the field on the electron shows an important difference between the optical and the electron lens. The potential V is continuously variable and hence would correspond to an optical lens with a continuously variable refractive index, but the optical lens in practice consists of several regions with a uniform refractive index with sharp boundaries between them.

A magnetic field can also be used to focus the electrons. To illustrate the action of the magnetic lens, we consider the simplified case in which the whole tube is in a region of uniform magnetic field. This could be produced by surrounding the tube with a solenoid. Let us suppose that the electrons are emitted from the source at O and consider an electron emitted with velocity v at an angle θ with the axis (Fig. 11-6). The focusing field is parallel to the electron beam, and the component of the velocity of the electron parallel to the field, $v \cos \theta$, produces a centripetal force which causes the electron to move in a circle normal to the field. The time taken to describe this circle is obtained from the gyrofrequency deduced earlier and is given by $2\pi m/eB$. Hence at a time $2\pi m/eB$ after emission, the electron has completed one circle and has returned to the axis. During this time it travels a distance $2\pi mv \cos \theta/eB$ along the axis, which for small θ is approximately $2\pi mv/eB$. Thus the focal length of the lens is given by

$$l = \frac{2\pi mv}{eB}.$$

Provided that the approximation $\cos \theta \simeq 1$ is valid, all the electrons are brought to a focus at this point. A more accurate expression for the focal length is

$$l = \frac{2\pi mv}{eB}\left(1 - \tfrac{1}{2}\theta^2\right),$$

and this formula can be used to obtain the spread of the image point due to the larger values of θ. The long focusing coil can be used in some applications, but in many cases, including, for example, the cathode-ray tube, it is inconvenient, and short coils are used. These will give nonuniform fields, and the calculations

*For a description of experimental mapping methods, see William H. Hayt, *Engineering Electromagnetics*, Chapter 6, New York: McGraw-Hill Book Co., 1958.

are much more complicated. In practice empirical methods will again be used. It is interesting to note that the magnetic lens is always a converging lens.

So far we have considered the action of the electric field and the magnetic field, each acting alone on the electron beam. We shall now discuss their combined action. Let us first consider an electron at rest at the origin and suppose that there is an electric field E in the $-x$-direction and a magnetic field B in the z-direction. Using the relations already developed, we can obtain the equations of motion of the electron. The electron will first begin to move in the x-direction under the influence of the electric field, but as soon as it acquires an x-velocity, there is a component of force in the y-direction owing to the action of the magnetic field. The force in the y-direction due to the magnetic field is eBv_x, where v_x is the x-component of the velocity of the electron. The electric field produces no force in the y-direction, and hence

$$m\,\frac{dv_y}{dt} = eBv_x.$$

For the x-component of the force we have eE due to the electric field and $-eBv_y$ due to the magnetic field; thus

$$m\,\frac{dv_x}{dt} = eE - eBv_y.$$

Hence

$$m\,\frac{d^2v_x}{dt^2} = -eB\,\frac{dv_y}{dt},$$

and substituting for dv_y/dt, we obtain

$$\frac{d^2v_x}{dt^2} = -\left(\frac{eB}{m}\right)^2 v_x = -\omega_0^2 v_x,$$

where $\omega_0 = eB/m$ is the gyrofrequency. At $t = 0$ the electron is at rest at the origin, and its acceleration is given by eE/m. Solving for v_x gives

$$v_x = A\sin\omega_0 t + C\cos\omega_0 t,$$

and the acceleration dv_x/dt is given by

$$\frac{dv_x}{dt} = \omega_0 A\cos\omega_0 t - \omega_0 C\sin\omega_0 t.$$

Substituting $dv_x/dt = eE/m$ at $t = 0$, we see that

$$\frac{eE}{m} = \frac{eB}{m}\,A \qquad \text{or} \qquad A = \frac{E}{B},$$

and therefore

$$v_x = \frac{E}{B}\sin\omega_0 t + C\cos\omega_0 t.$$

But since $v_x = 0$ at $t = 0$, we have $C = 0$, and therefore

$$v_x = \frac{E}{B} \sin \omega_0 t.$$

Integrating again, we obtain

$$x = -\frac{E}{\omega_0 B} \cos \omega_0 t + D.$$

But $x = 0$ at $t = 0$; thus

$$D = \frac{E}{\omega_0 B},$$

and hence

$$x = \frac{E}{\omega_0 B} (1 - \cos \omega_0 t).$$

To obtain y we have

$$v_y = \frac{E}{B} - \frac{1}{\omega_0} \frac{dv_x}{dt} = \frac{E}{B} - \frac{E}{B} \cos \omega_0 t,$$

$$y = \frac{E}{B} t - \frac{E}{\omega_0 B} \sin \omega_0 t,$$

or

$$y = \frac{E}{\omega_0 B} (\omega_0 t - \sin \omega_0 t),$$

the integration constant being zero since $y = 0$ at $t = 0$.

The equations for the trajectory in parametric form are therefore

$$x = \frac{E}{\omega_0 B} (1 - \cos \omega_0 t), \qquad y = \frac{E}{\omega_0 B} (\omega_0 t - \sin \omega_0 t),$$

which are the equations of a cycloid. Let us suppose that the electron is in a beam with a steady velocity in the z-direction. The fields produce no forces

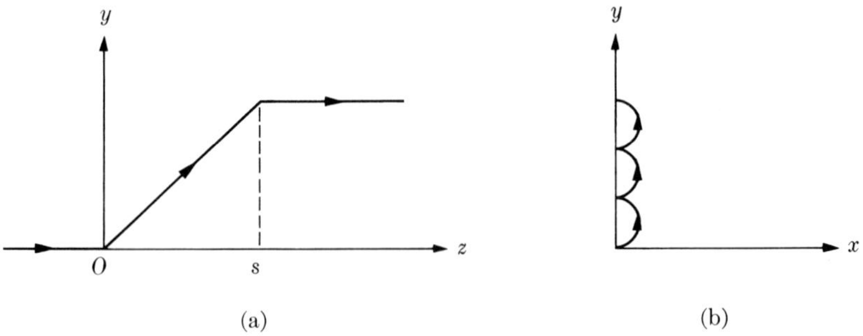

FIG. 11–7. The orthicon deflection system.

in the z-direction, and so this steady velocity is not disturbed. We next suppose that the electric field is produced by deflector plates confined to the region $0 < z < s$. The magnetic field is in the same direction as the beam, and hence will produce no deflection of the electrons acting by itself. Thus there is no reason to confine the B-field. (It would, in fact, produce some focusing action, as we have seen.) Hence the beam travels in the z-direction for $z < 0$. For $0 < z < s$, its trajectory projected on the xy-plane is a cycloid, and its steady component of velocity is still in the z-direction (Fig. 11-7). If the electron velocity is v, then the time during which it is under the influence of the electric field is s/v. By suitably adjusting the accelerating voltage (which determines v), the length of the deflector plate s, and the magnetic field B, we can make $eBs/mv = 2n\pi$, that is, in the above formula $\omega_0 t = 2n\pi$. Hence we see that

$$x = 0, \qquad y = \frac{mE}{eB^2} 2n\pi,$$

$$v_x = 0, \qquad v_y = 0.$$

Under these conditions, the deflection in the y-direction is proportional to the electric field E. The beam emerges from the deflector assembly with its original velocity in the z-direction. We have therefore achieved a parallel deflection of the beam. The amount of the deflection can be controlled by the electric field, and the direction of the emerging beam is independent of the deflection. This type of deflection system is used in the orthicon type of camera tube, where the beam is required to strike the screen perpendicularly regardless of the deflection during the scanning process.

11-3 The mass spectrograph

It is of interest to measure the masses of the atomic particles and to be able to analyze given samples of gas to determine their composition. We have seen above that the mass of the particle enters into the expressions for the deflection of beams, and thus there is the possibility that this effect can be used to measure m. Before going into the details of designing the fields, we shall first consider the general features of the mass spectrograph, the instrument which has been developed for this purpose, and some of the problems that the early workers had to solve.

To deflect the particles by means of fields requires that the particles carry an electric charge. The study of electrical discharges in gases revealed the existence of the "positive rays"—streams of particles carrying positive charges. Thus the particles can be given an electric charge by means of an electrical discharge in the gas. Such a discharge, however, can only be obtained at pressures which are relatively high compared to the pressure customarily used in vacuum and cathode-ray tubes. The controlled deflection of the beam in a cathode-ray tube obviously requires that the electrons shall not be deflected by collisions

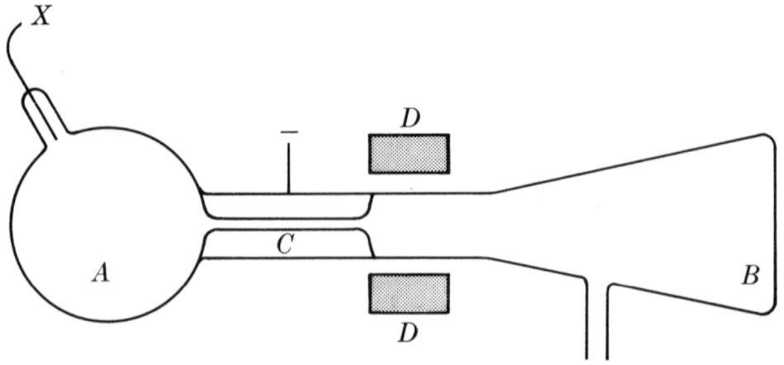

FIG. 11-8. The mass spectrograph.

with gas molecules between the deflector assembly and the screen. At the relatively high pressures needed for the electric discharge, this condition is not fulfilled. Although the positive rays are much more massive, since they consist of atoms rather than electrons, collisions with neutral molecules will still deflect them and create confusion. This is made worse by the fact that some of these collisions will be ionizing collisions. The charge on the particle may therefore change on its way to the screen. These problems are overcome in the apparatus developed by J. J. Thompson (Fig. 11-8). The apparatus consists of two bulbs, A and B, separated by an iron tube C. The gas under investigation is contained in A and is ionized by an electrical discharge between the two electrodes. The tube C forms the cathode, and the positively charged particles are attracted to it. Some of these pass down the fine capillary tube in the center of C and are formed into a narrow beam of positive particles. The iron material of the tube serves to shield these particles from stray magnetic fields. The bulb B is continuously pumped through the side tube and is thus maintained at a much lower pressure than A, since the only connection is via the fine tube in C. This incidentally is an interesting pneumatic circuit. We maintain a pressure difference (voltage) by means of a gas flow (current) through a capillary tube (resistance). The resistance to the air flow arises from the viscosity of the gas. The tube C is thus made to serve the threefold purpose of a cathode, a magnetic shield, and a pneumatic resistance. The beam is deflected by the deflector assembly D, which produces both electric and magnetic fields. We shall now calculate these deflections.

The fields employed in the equipment are parallel so that the electric field produces a deflection in the x-direction, say, and the magnetic field in the y-direction. From the formulas of Section 11-2, we have, for the electric deflection,

$$x = l_1 s_1 \frac{e}{m} \frac{E}{v^2},$$

and for the magnetic deflection,

$$y = l_2 s_2 \frac{e}{m} \frac{B}{v}.$$

In these formulas l_1, s_1, l_2, s_2, are dimensions of the tube and will be constant. For each type of particle, the mass m will be constant, and supposing that they are all singly ionized (i.e., have lost one electron), then e will be a constant. We can maintain the deflecting fields constant, and hence the only variable will be v, and v will depend on the point in the bulb A at which the particle acquired its charge, since this will determine the electric potential through which it falls before reaching C. Therefore eliminating v between these two equations, we have $y^2/x = \text{const}$, which is the equation of a parabola. Hence all particles with the same value of e/m will produce a spot on a parabola regardless of their velocity. There is a minimum value of deflection given by the maximum value of v, which occurs when the particle acquires its charge near the anode and falls through the whole potential difference V between anode and cathode. The x-deflection in this case is obtained by substituting $\frac{1}{2}mv^2 = eV$, and so $x_{\min} = \frac{1}{2}l_1s_1E/V$. This is independent of e/m and thus is the same for all particles. These parabolas can be recorded by replacing the fluorescent screen of the cathode-ray tube by a photographic plate. The resulting patterns take on the form shown in Fig. 11-9. The different parabolas correspond to different values of e/m. The part of the curve below the x-axis arises because the particle may lose its positive charge and even acquire a negative one as it passes down the capillary. The negative charge gives the lower half of these curves.

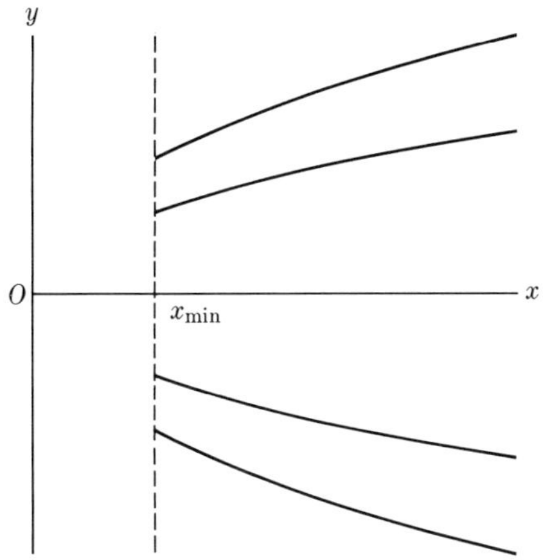

FIG. 11-9. A mass spectrum.

Note also that the equipment described above essentially measures e/m. We cannot always identify the chemical atom responsible for a given trace. For example, carbon monoxide consisting of one carbon atom of mass 12 and one oxygen atom of mass 16 has molecular mass of 28, which is the same as that of the nitrogen molecule consisting of two nitrogen atoms of mass 14. Furthermore, if the carbon monoxide molecule carries a double charge and a nitrogen atom, which has half the mass, carries a single charge, then both have the same value of e/m and fall on the same parabolas. The curves must therefore be interpreted with some care. An important difference between this method of measuring mass and chemical methods is that here we measure the mass of the individual atoms and not the average mass taken over a large number of atoms. This has led to the discovery of many isotopes—atoms with the same chemical properties but different masses. For example, oxygen has three isotopes with masses of 16, 17, and 18.

We have seen that the different velocities of particles with the same e/m produce a parabola. There is no advantage in having the particles spread in this way. In fact, it is a disadvantage. If we could arrange it so that all the particles with the same e/m strike the same spot regardless of velocity, we would increase the sensitivity of the instrument. In other words, we would like to introduce some focusing. Aston designed a mass spectrometer in which this was accomplished as follows. The beam is first passed through an electric field which produces an angular deviation of the beam θ_E. Using the formula from Section 11–2, we can find θ_E:

$$y = ls\frac{e}{m}\frac{E}{v^2},$$

$$\therefore\ \theta_E = \frac{y}{l} = s\frac{e}{m}\frac{E}{v^2}.$$

Thus the electric field E spreads the particles into a spectrum, and the angular deviation depends on e/m and v. For a constant electric field and a given e/m in a particular tube (s constant), we have $\theta_E v^2 = \text{const.}$ Hence by differentiating we obtain

$$v^2\,d\theta_E + 2v\theta_E\,dv = 0 \qquad \text{or} \qquad \frac{d\theta_E}{\theta_E} + 2\frac{dv}{v} = 0.$$

This formula gives the difference in deviation $d\theta_E$ for two particles with the same value of e/m, but velocities differing by dv. The beam is now passed through a magnetic deflecting field which gives an angular deviation θ_M. From Section 11–2

$$\theta_M = s'\frac{e}{m}\frac{B}{v},$$

and in this case we have $\theta_M v = \text{const.}$ Hence we see that

$$\frac{d\theta_M}{\theta_M} + \frac{dv}{v} = 0.$$

If we arrange for $d\theta_M$ and $d\theta_E$ to cancel, the net difference in the sum of the deviations due to the electric and magnetic fields is zero, and particles with the same e/m emerge from the deflection system with parallel trajectories regardless of differences in velocity. Putting $d\theta_M = -d\theta_E$, we have

$$\theta_M\frac{dv}{v} = -2\theta_E\frac{dv}{v} \qquad \text{or} \qquad \theta_M = -2\theta_E.$$

Hence to produce a parallel beam the electric and magnetic fields are adjusted so that the magnetic deflection is in the opposite direction to the electric deflection and has twice the magnitude. We should check to see that we have not also destroyed the separation of the particles in accordance with their values of e/m. This is not in fact the case, and the difference arises because both deflections are proportional to e/m, whereas the velocity dependence is $1/v^2$ for the electric and $1/v$ for the magnetic case.

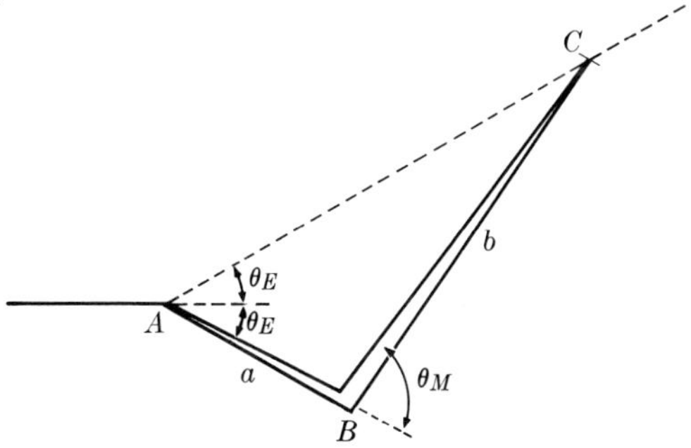

FIG. 11-10. The focusing mechanism for the mass spectrograph.

The above arrangement of the fields produces a parallel beam. Actually we would like to go a stage further and produce a converging beam which comes to a focus. We can do this by increasing the magnetic deflection. Let us suppose that the beam passes through the electric deflector assembly at A (Fig. 11-10), travels a distance a to the magnetic deflector assembly at B, and then goes a distance b to the focus at C. The spread of the beam due to the electric deflection is $d\theta_E$ times the total distance traveled. Hence at C the electric deflection produces a spread $(a + b)\, d\theta_E$. Similarly, the spread of the beam at C due to the magnetic deflection is $bd\theta_M$. If these are to cancel, $(a + b)\, d\theta_E = -bd\theta_M$. The negative sign in the expression above means that the deflection must be in opposite senses for the two fields. This has already been taken into account in Fig. 11-10 by the manner in which the figure has been drawn. Hence, from the expressions above for $d\theta_E$ and $d\theta_M$, we have

$$(a + b)2\theta_E = -b\theta_M.$$

Note that if $\theta_M = -2\theta_E$, the above formula requires $1 + a/b = 1$, which if $a \neq 0$ means that b is infinite. This checks since the condition $\theta_M = -2\theta_E$ gives a parallel beam. From the relation above we obtain

$$\frac{a}{b} = \frac{\theta_M - 2\theta_E}{2\theta_E},$$

which is satisfied in the triangle ABC if $\angle BAC$ is $2\theta_E$. We then have $\angle ACB = \theta_M - 2\theta_E$, and since all these angles are small, so that we may replace the sines of the angles by the angles themselves, the above equation follows from the geometry of the triangle. Hence it is seen that the focus falls on a line drawn through the electric deflector assembly and making an angle to the undeflected beam equal in magnitude and opposite in sense to the angle of deflection due to the electric field.

In practice most of the particles acquire a kinetic energy roughly equal to eV_1, that is, in the parabolic plots the majority of the particles are clustered near $x_{\min}$. In this case we know from the formula in Section 11-2 that

$$\theta_E \simeq \frac{Es}{2V},$$

which is independent of e/m. Thus particles of all e/m have the same average deflection, and the spread in velocities gives the beam a finite width or divergence. This obviously simplifies the design of the instrument. The focal points for all e/m lie on the line AC, and a photographic plate is placed along this line to record the spectra. The value of e/m can be calculated, but it is common practice to introduce some atoms of known mass to obtain an empirical calibration of the instrument.

11-4 Nuclear accelerators

Thus far we have considered applications of the laws we have developed to laboratory instruments such as the cathode-ray tube and the mass spectrograph. These same laws are equally applicable to large electrical machinery, and we will discuss the nuclear accelerators as an example.

We saw that an electron falling through a potential of 1 v acquires an energy of 1 eV. This is the simplest method of accelerating charged particles and was the earliest method used. High-voltage electric generators such as those developed by Cockcroft and Walton and by van der Graff were employed to produce intense electric fields to accelerate the particles. However, because of the difficulties in working with high voltages, it was found that these methods could be used only to produce particles with energies of a few million electron volts (MeV). Since these energies are not high enough for many of the purposes of the nuclear physicist, other methods of accelerating particles were sought.

The cyclotron, developed by Lawrence, was the first of the accelerators not relying on high voltages. The principles on which it works are those we have discussed earlier in the chapter. In this machine the particles are accelerated by a relatively low voltage between two dees, as shown in Fig. 11-11. The dees are enclosed in a vacuum chamber in which the particles to be accelerated are made to travel in circular paths by a magnetic field. The voltage across the dees is an oscillating voltage, and the frequency is adjusted so that this voltage is reversed in the time it takes for the particle to complete a half-revolution. Thus each time the particle reaches the gap between the dees it is accelerated. We have seen that a particle with charge e moving perpendicular to a magnetic field B with velocity v executes a circle of radius $r = mv/eB$ in a time $T = 2\pi m/eB$. The cyclotron is based on the fact that this time is independent of the energy of the particles. Thus, although the particles are being accelerated and their energy is increasing, they still take the same time to complete the half-circle between successive crossings of the gap between the dees. Hence an oscillator of constant frequency $\omega = eB/m$ produces an electric field across the gap which always accelerates the particles. The trajectory of the particles consists of a series of semicircles of increasing radius, and the curve resembles a spiral (Fig. 11-11). Eventually the particle reaches the outer edge of the dees, and the beam is made to leave the accelerator by means of a deflector plate and can be used to perform experiments. If the radius of the dees is r, then the energy of the emerging particle is $(eBr)^2/2m$. The machine will produce energies up to 10 MeV. At this energy, difficulties are experienced because the particles are now traveling at speeds where the increase in mass due to relativistic effects is no longer negligible. Thus the period of revolution of the particle which depends on m is no longer constant, and synchronism with the accelerating fields is lost.

The betatron, an electron accelerator developed by Kerst, is another machine not requiring high voltage. A simplified diagram of the betatron is shown in Fig. 11-12. Let us suppose that we have electrons moving in a circle. These constitute a current in a closed circuit. If we change the magnetic flux linked with this circuit, there will be an electromotive force acting in the circuit which, with the signs suitably chosen, will cause an increase in the current, i.e., the electrons will be accelerated. This is the principle of the betatron. So long as the current flows in a wire, there are no difficulties. However, if the electrons are flowing in a vacuum as they must be in particle accelerators, then the fields must be carefully designed to keep the electrons in a proper orbit. Our first requirement is that the electrons move in a circle of constant radius all the time they are being accelerated. To accomplish this we need a magnetic field perpendicular to the plane of the orbit. We suppose that the orbit has

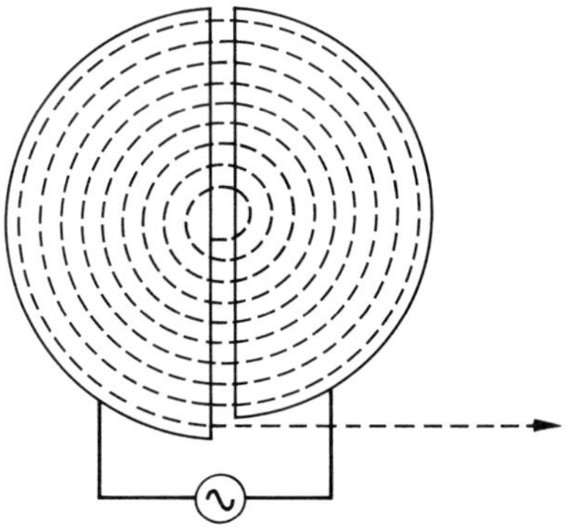

FIG. 11-11. The cyclotron.

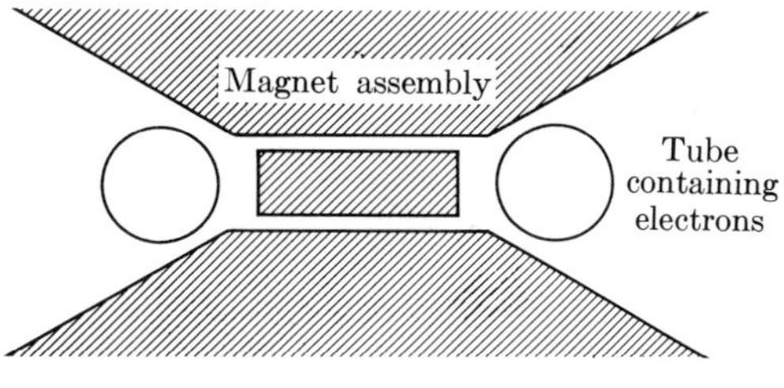

FIG. 11-12. The betatron.

radius r_0; then the required value of the magnetic field is $B_0 = mv/er_0$. So long as B has the value B_0 at $r = r_0$, the electron will stay in its circular orbit, and we are free to adjust B at the other values of r to suit other requirements. If we change B inside the orbit, we have changed the flux linked with the circuit, and v will change, which in turn will require a change of B_0 to maintain the orbit. It is therefore convenient to divide the field into two parts, and indeed, the magnetic system of the betatron consists of an accelerating field in the center and a guiding magnetic field surrounding it. A doughnut-shaped vacuum tube is situated in the annular guiding field to contain the electrons. We now calculate how these fields are related to each other.

From Section 5–2 we know that

$$-\int \mathbf{E} \cdot d\mathbf{l} = \frac{\partial \psi}{\partial t}.$$

Let us suppose that the magnetic field has an axial symmetry about the z-axis. The electric field produced by changing this field must have a zero divergence, since no electric charge is involved, and hence the **E**-lines must be closed curves. Further, from the symmetry of the case, the **E**-field can depend only on the distance from the z-axis, and thus the **E**-lines are circles about the z-axis (cf. Section 6–8). Hence, if we calculate the integral around the circle radius r_0, we get the value $2\pi r_0 E$, which must be equal to the rate of change of total flux through the circle radius r_0. The force acting on an electron in this orbit is eE, and this is equal to mass times acceleration. We therefore obtain the rate of change of velocity as

$$m\frac{dv}{dt} = eE = \frac{e}{2\pi r_0}\frac{d\psi}{dt},$$

where ψ refers to the total flux through the orbit. We can relate this to the magnetic field at $r = r_0$ by using the relation $mv = er_0B_0$, and thus

$$m\frac{dv}{dt} = er_0\frac{dB_0}{dt},$$

since we have specified $r_0 = \text{const}$. Hence we have

$$r_0\frac{dB_0}{dt} = \frac{1}{2\pi r_0}\frac{d\psi}{dt}.$$

But the average magnetic field inside the orbit $B_{av} = \psi/\pi r_0^2$, and thus

$$\frac{dB_0}{dt} = \frac{1}{2}\frac{dB_{av}}{dt}.$$

To keep the electron in a constant radius orbit the fields must be so designed that the rate of increase of field at the radius of the orbit is half the rate of increase of the average field over the whole orbit. We see that B_0 is determined completely from the relation $mv = B_0 e r_0$, and as v increases B_0 must increase. But B_{av} is determined only by its rate of change, and hence a constant field can be added to it. Thus if B_0 varies from 0 to 1000, B_{av} could vary from 0 to 2000, or alternatively a biasing field of 1000 could be added so that B_{av} varies from -1000 to $+1000$. This scheme has its advantages; machines employing it are known as biased betatrons.

The magnetic fields are usually varied sinusoidally at 60 cycles/sec, so that the standard power supply can be used to drive the magnets. The electrons are injected into the machine at an energy of 50 kiloelectron volts (keV) from an electron gun at the appropriate point of the magnetic field cycle and are taken out when their energy reaches a peak. Although a small fraction of the electron beam can be brought out of the machine, the output is usually in the form of x-rays. A sudden reduction of the magnetic field produced by an independent coil causes the orbits to contract, and the electrons strike a tungsten target and produce x-rays. The betatron can produce energies up to 100 MeV. At this stage the electrons lose a significant amount of energy by radiation due to their centripetal acceleration. (We saw in Chapter 9 that a time-varying current, i.e., accelerating charge, produces electromagnetic radiation.) This alters the orbit radius unless additional flux is added to compensate for the loss. Adding flux leads to considerable complication, and therefore other machines are used for higher energies.

Returning now to a consideration of the orbit, we note that for the machine to operate successfully the orbit must be stable; i.e., if the electron receives some small perturbation causing it to move out of orbit, it must be acted on by restoring forces which return it to its original orbit. This is accomplished by appropriate design of the guiding fields. We remarked above that the field could be a function of distance from the center of the orbit, and so far the only restriction placed on this is that the rate of change of the average field must be twice that of the field at the orbit radius. We can satisfy this requirement by suitable adjustment of the accelerating field near the center of the machine, leaving the detailed variation of the guiding field, i.e., the field near $r = r_0$, to be adjusted to achieve orbit stability. We must consider perturbations in the radial and in the axial directions.

First we will discuss the radial stability. The momentum of the electron is given by $mv = eBr$. Thus we see that if B is made to vary like $1/r$, the momentum will be independent of r. Hence the electron could move from an orbit radius r_0 to one of radius $r_0 + dr$ in a finite time dt, and its rate of change of momentum would be zero. Hence there would be no force acting on it, and it would be free to drift into such an orbit. If the variation of the field is different from $1/r$, then clearly a rate of change of momentum would be involved, and forces would be brought into play. The magnetic field responsible for the force is the difference between the actual field and the $1/r$ field. For example, Fig. 11–13 shows two fields, both of which have the value B_0 at r_0, as required to produce the correct orbit radius. One of these fields, B_1, varies like $1/r$, and

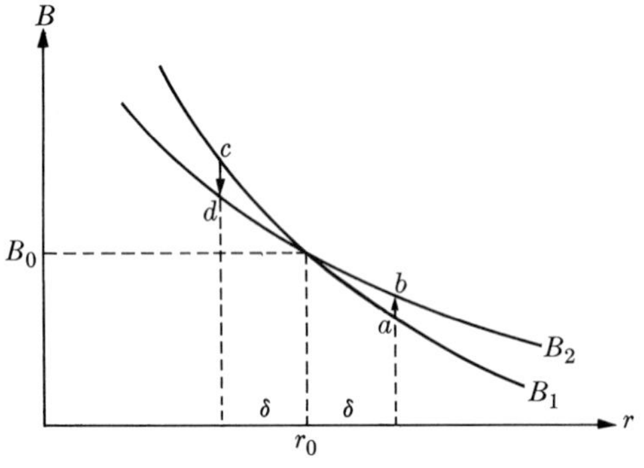

FIG. 11-13. The condition for radial stability of the betatron orbit.

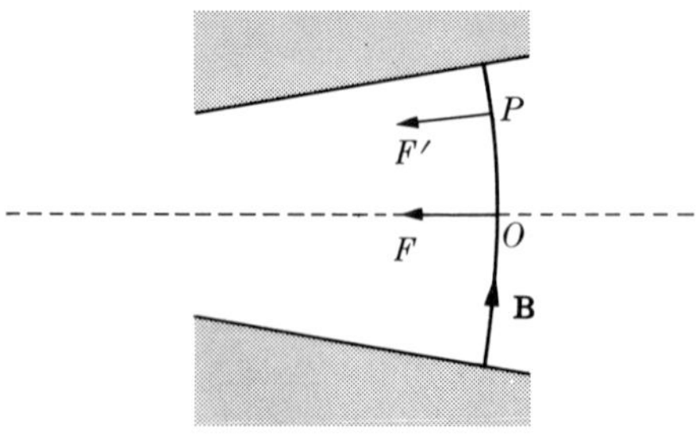

FIG. 11-14. The condition for axial stability of the betatron orbit.

the other varies more slowly. At $r = r_0 + \delta$, B_2 is larger than B_1, and thus we have a difference field *ab* to produce a force on the electron. This is in the same direction as B_0 and will exert a force on the electron in the same direction as the force due to B_0, that is, toward the center. At $r_0 - \delta$ we have the difference field *cd*, which is in the opposite direction to B_0 and will therefore produce a force in the opposite direction, i.e., away from the center. Thus, if the field in the vicinity of r_0 varies like B_2 (more slowly than $1/r$), then it produces a stabilizing or focusing action.

The focusing action in the axial direction is relatively easy and may be explained by reference to Fig. 11-14. The field lines are shaped as shown, the center of the machine being to the left. To keep the particle in orbit the force due to line B must be in the direction F. If the particle drifts to a point P, the force is in the direction F' perpendicular to the direction of $\mathbf{B}$. This force is seen to have a component acting toward O and so returning the particle to its original orbit. Thus by suitable design of the guiding field the appropriate focusing actions are obtained, and a stable orbit is achieved.

There are several other types of particle accelerators in existence which have been designed to produce particles of higher and higher energy. Some of these machines are very large and costly. The proton synchrotron (or bevatron) has been built with an orbit radius of some 50 ft and requires a magnet weighing thousands of tons. Machines of this type are capable of producing energies of 10 BeV (10^{10} eV).

11-5 The diode

In the calculations made in this chapter, we have neglected the fields of the charged particles themselves. Thus we have assumed that each particle moves independently of all the others, or in other words, that the forces between particles can be neglected. We have accordingly not taken into account the *space charge effects,* as they are customarily called. In studying diodes, we are used to dividing their operation into two regions, the space charge-limited region and the temperature-limited region. A study of the first region enables us to carry our theory a stage further and include the effect of the forces between particles, the space charge effects.

Let us consider a diode consisting of two parallel planes, one forming the cathode and the other the plate. The electrons are emitted from the cathode and travel toward the plate under the influence of the field arising from the positive potential on the plate. If the supply of electrons from the cathode is large enough compared with the field, then the cathode is supplying electrons faster than they can be carried away and collected by the plate. Hence the electron concentration builds up near the cathode, and a negative space charge develops. We can see that such a situation is possible if we go to the limit and make the plate potential zero so that there is no field at all to draw the electrons away. The space charge produces a field at the cathode in opposition to that produced by the plate voltage, and this tends to oppose the emission of the electrons from the cathode surface. When the space charge has built up to the point where the number of electrons emitted by the cathode is equal to the number collected by the plate, an equilibrium is established. Since the electrons are emitted from the cathode with a finite velocity, a reverse field would be required to stop the emission altogether. However, to obtain equilibrium we do not wish to stop the emission completely, since the emission from the cathode must replace those electrons collected by the plate. We usually assume that the field at the cathode is zero at equilibrium, which is a good approximation in practice. We shall now carry through the calculation for these conditions.

Let us consider the diode shown in Fig. 11–15. We will suppose that the plates are infinite so that the solution is independent of y and z and we have a one-dimensional problem. At equilibrium none of the variables will be functions of time (by definition). Hence, from the equation of continuity, we obtain $\operatorname{div} \mathbf{J} = -\partial q_v/\partial t = 0$, and thus $\partial J_x/\partial x = 0$ and $J_x = \text{const}$. In this problem the current will be flowing in the x-direction, and so we drop the suffix and write $J = \text{const}$. The potential V in the region between the plates, not being a function of time, will be the electrostatic potential and will therefore satisfy Poisson's equation:

$$\frac{d^2V}{dx^2} = -\frac{q_v}{\epsilon_0}.$$

The velocity of the electron is given by the relation from Section 11–1,

$$v = \sqrt{2eV/m},$$

and we now have two equations among the unknowns V, q_v, and v (all of which are functions of x). Therefore we need one more to solve the problem, and this we obtain from a general relation between q_v, v, and J. We know that a moving

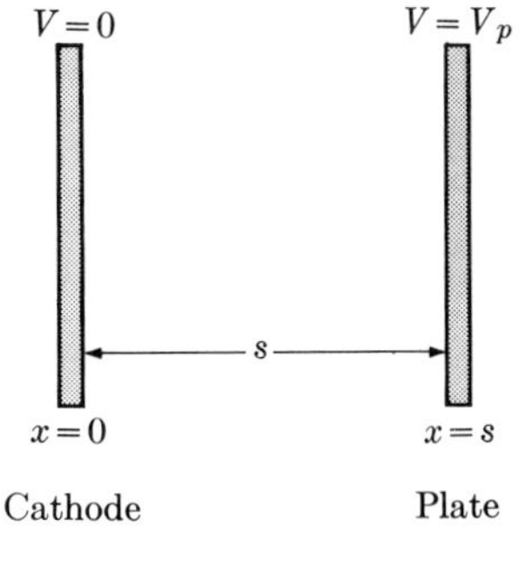

FIG. 11–15. The plane diode.

charge is equivalent to a current. Let us consider an element of area ΔA perpendicular to the velocity vector $\mathbf{v}$. In time dt the charge moves a distance $v\,dt$, and, since ΔA is perpendicular to $\mathbf{v}$, $v\,\Delta A\,dt$ represents the volume shown in Fig. 11–16. If the charge density is q_v, then the charge in this volume is $q_v v\,\Delta A\,dt$, all of which entered the volume by the face ΔA. Hence the charge crossing ΔA per unit time is $q_v v\,\Delta A$. However, this is just the current through ΔA, and thus the current density is given by $J = q_v v$. Since $\mathbf{J}$ and $\mathbf{v}$ are in the same direction, we have the vector equation

$$\mathbf{J} = q_v \mathbf{v},$$

and since in this problem $\mathbf{J}$ and $\mathbf{v}$ have only x-components, we revert to the scalar form. This provides the third equation needed. To obtain the relation between the current through the diode and the plate voltage we must relate J and V. We eliminate q_v and v from these equations and arrive at

$$\frac{d^2V}{dx^2} = -\frac{J}{\epsilon_0 v} = -\frac{J\sqrt{m/2e}}{\epsilon_0\sqrt{V}} = -k\frac{1}{\sqrt{V}},$$

where k is a constant independent of x. The standard method of integrating this equation is to multiply by $2\,dV/dx$:

$$\frac{d}{dx}\left(\frac{dV}{dx}\right)^2 = -2k\frac{1}{\sqrt{V}}\frac{dV}{dx} = -4k\frac{d}{dx}\sqrt{V}.$$

Integrating, we obtain

$$\left(\frac{dV}{dx}\right)^2 = -4k\sqrt{V} + \text{const.}$$

From the above discussion, it follows that at the cathode, $x = 0$, $V = 0$, and $dV/dx = 0$, so that the constant of integration is zero. We therefore have

$$\frac{dV}{V^{1/4}} = \sqrt{-4k}\,dx,$$

and by integrating again we see that

$$(4/3)V^{3/4} = \sqrt{-4k}\,x + \text{const},$$

and again the constant of integration is zero since $V = 0$ for $x = 0$. Hence

$$(16/9)V^{3/2} = -4(J/\epsilon_0)\sqrt{m/2e}\,x^2,$$

and since $V = V_p$ for $x = s$, we finally have

$$J = -(4/9)(\epsilon_0/s^2)\sqrt{2e/m}\,V_p^{3/2}.$$

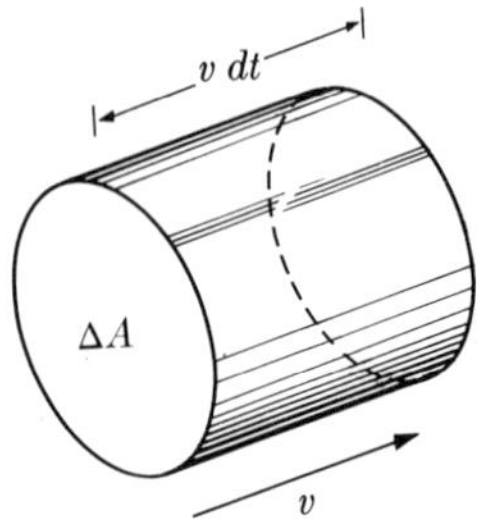

FIG. 11–16. To obtain the relation between the charge density, the current density, and the electron velocity.

Note that the current is negative as it must be since a stream of electrons moving in the $+x$-direction constitutes a current in the $-x$-direction. Also the current is proportional to $V_p^{3/2}$, which gives us a nonlinear relation between current and voltage. This equation is referred to as the *three-halves power law* and also as the *Child-Langmuir relation.*

Writing V as a function of x and expressing J in terms of V_p gives

$$V = V_p \left(\frac{x}{s}\right)^{4/3}.$$

Note that in the absence of the space charge we would have $V = V_p(x/s)$. The field is readily obtained from the expression

$$E = -\frac{dV}{dx} = -\frac{4V_p}{3s^{4/3}} x^{1/3},$$

and the space charge density is obtained from Poisson's equation:

$$q_v = -\epsilon_0 \frac{d^2V}{dx^2} = \epsilon_0 \frac{dE}{dx} = -\frac{4\epsilon_0 V_p}{9s^{3/4}} x^{-2/3}.$$

Note that the space charge is negative, which again is as it must be since the particles involved are electrons. We also note that this relation gives an infinite space charge density at the origin (i.e., the cathode surface). This is because of our assumption that the field is zero at the cathode. In practice it will have a small but finite value.

Problems

11–1 The suppressor grid of a vacuum tube is to prevent secondary electrons emitted from the plate from being collected by the screen grid. Assuming a plane construction for simplicity, what is the maximum velocity which the emitted electron can have if it is to be returned to the plate, given that the suppressor grid is grounded and the plate is at voltage V_p?

11–2 The inner surface of one of the plates of a parallel-plate capacitor is coated with a photoemissive material. If the maximum velocity of the emitted photoelectrons is v_m, what value of magnetic field parallel to the plates is required to prevent the electrons from reaching the other plate?

11–3 Derive an equation for the trajectory of the electron in the region of the deflector plates of the electrostatic deflection system. What is this curve?

11–4 The deflector plates of a cathode-ray tube are 2 cm long, are spaced 1 cm apart, and are placed at a distance of 20 cm from the screen. If the accelerating voltage is 1000, calculate the deflection sensitivity. What is the deflection for 100 v applied across the plates?

11–5 Investigate the problem of high-frequency cutoff in the magnetic deflection system by considering a square-wave input.

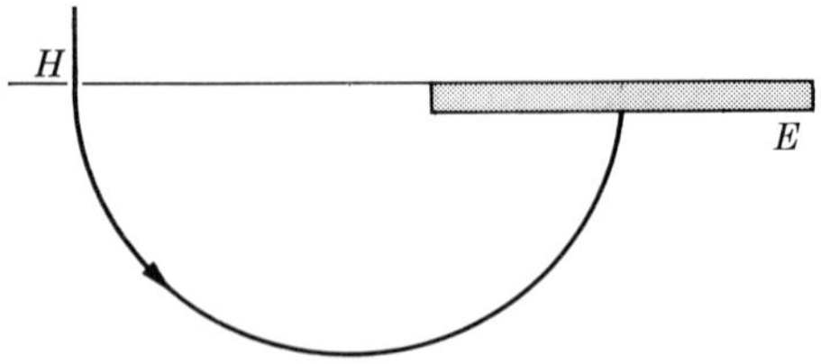

FIG. 11–17. A mass spectrometer.

11–6 Calculate the magnitude of the magnetic field required to deflect the beam of a cathode-ray tube 10 cm, given that the deflector coils have a radius of 2 cm, are at a distance of 20 cm from the screen, and the accelerating voltage is 1000.

11–7 To obtain an order of magnitude, use the formula for the magnetic field at the center of a circular current to determine the number of turns in the coil that will produce the field required in Problem 11–6, given that the source driving the deflector coils can produce 500 ma.

11–8 Calculate the field required to give a focal length of 20 cm for a magnetic focusing system, using a value for the velocity corresponding to an accelerating voltage of 1000.

11–9 Calculate the magnetic field required to produce 10-MeV protons in a cyclotron with dees of radius 2 m.

11–10 Assuming that the B-lines in Fig. 11–14 are circular, calculate the restoring force on the electron as a function of displacement from its equilibrium orbit. How does this force vary as the particle is accelerated?

11–11 A mass spectrometer is designed so that the ions enter through a small hole H in a plate P and are then deflected through 180° by a magnetic field and recorded on a photographic plate E (Fig. 11–17). Investigate how the position of the point at which the electrons strike the plate varies when the particles enter at a small angle to the normal to the plate. Explain how this device has some focusing properties.

11–12 If a small alternating voltage is superimposed on the steady voltage across the diode of Section 11–5, how is the steady current through the device affected? [*Hint:* Consider J^2 and retain up to second-order small quantities.]

11–13 For the diode of Section 11–5, consider plates 2 cm square separated by a distance of 2 cm and with 100 v across them. Calculate the current through the tube.

CHAPTER 12

Ionized Gases and Ferrites

12–1 Introduction

In the last chapter we studied the motion of charged particles under the influence of fields. The formulas derived were developed from the relations which apply when the fields are static. Although the fields were allowed to vary with time, it was implicitly assumed that the variation was "slow" so that the static formulas still applied. Such conditions are termed *quasistatic*. In this chapter, following the development we made in the absence of charged particles, we shall study the interaction of charged particles and fields when the fields are rapidly alternating. We shall ignore space charge effects as we did in the last chapter (except for the final section). In the second part of the chapter, we shall study some of the properties of the ferrites which produce results that are in many ways similar to those of ionized gases.

12–2 Constitutive relations

The ionized gas in the present problem can be considered to be the medium in which the fields exist. We saw in Section 5–5 that the effects of the medium on the fields enter the equations via the constitutive relations. We shall accordingly first derive the constitutive relations for the ionized gas which we have, in fact, already done in Section 8–2. We saw there that it was convenient to use the relation $\mathbf{D} = \epsilon_0\mathbf{E} + \mathbf{P}$ and to work in terms of the susceptibility. We shall follow a similar calculation for a more complicated case. We suppose that in addition to the ionization, there is a uniform magnetostatic field. These two, ionization plus static field, comprise the medium which is referred to as magneto-ionic.

We showed in Section 8–2 that if the electron is displaced a distance x from its equilibrium position, it produces a polarization Nex. Polarization is a vector, and the direction of the dipole moment is considered to be the direction from the positive to the negative charge. Hence the direction of the polarization is the same as the direction of the displacement. For displacement in a general direction $\mathbf{l}$, we would have $\mathbf{P} = Ne\mathbf{l}$.

We calculated the displacement produced by a sinusoidally time-varying electric field from the equation of motion. The case under consideration here is complicated by the fact that as soon as the electron acquires a velocity, another force, $e\mathbf{v} \times \mathbf{B}$, comes into play owing to the presence of the magnetic field. At this point it is convenient to make a special choice of axes so that the formulas do not become unnecessarily cumbersome. Later we shall consider waves propagating in the medium, and following our previous practice, we shall take the z-direction as the direction of propagation. The simplest case to investigate is that of a wave propagating in the same direction as the steady magnetic field. We shall accordingly derive the constitutive relations for the field in the z-direction.

The equation of motion of an electron with velocity $\mathbf{v}$ in the presence of electric and magnetic fields is

$$m\frac{d\mathbf{v}}{dt} = e\mathbf{E} + e\mathbf{v} \times \mathbf{B},$$

which contains the product term $\mathbf{v} \times \mathbf{B}$ and so is nonlinear. Since it is much more difficult to deal with nonlinear equations than with linear ones, we shall linearize the equation by using the *small-signal approximation.* In considering wave propagation we are interested in the ac-behavior of the medium. Previously, when investigating waves, we assumed that all our variables vary sinusoidally with time. However, here we have specifically introduced a steady field as part of the medium, and so this assumption can no longer be made. Thus the magnetic field will consist of a dc-component (from the medium) and an ac-component from the wave, and we write

$$\mathbf{B} = \mathbf{B}_0 + \operatorname{Re}(\mathcal{B}e^{j\omega t}).$$

The small-signal approximation is based on the assumption that the magnitude of $\mathcal{B}$ is small compared to the magnitude of $\mathbf{B}_0$. Since this approximation is of considerable importance, we shall investigate how it simplifies the general equation

$$\mathbf{A} = \mathbf{B} \times \mathbf{C}.$$

In the general case both $\mathbf{B}$ and $\mathbf{C}$ consist of a dc- and an ac-term. In the product we can neglect the product of ac-quantities, these being second-order small quantities, and hence

$$\mathbf{A}_0 + \operatorname{Re}(\mathcal{A}e^{j\omega t}) = \mathbf{B}_0 \times \mathbf{C}_0 + \mathbf{B}_0 \times \operatorname{Re}(\mathcal{C}e^{j\omega t}) + \operatorname{Re}(\mathcal{B}e^{j\omega t}) \times \mathbf{C}_0.$$

But this must be true for all t. Putting $t = 0$ gives

$$\mathbf{A}_0 + \operatorname{Re}(\mathcal{A}) = \mathbf{B}_0 \times \mathbf{C}_0 + \mathbf{B}_0 \times \operatorname{Re}(\mathcal{C}) + \operatorname{Re}(\mathcal{B}) \times \mathbf{C}_0,$$

and putting $\omega t = \pi$ yields

$$\mathbf{A}_0 - \operatorname{Re}(\mathcal{A}) = \mathbf{B}_0 \times \mathbf{C}_0 - \mathbf{B}_0 \times \operatorname{Re}(\mathcal{C}) - \operatorname{Re}(\mathcal{B}) \times \mathbf{C}_0,$$

from which we have

$$\mathbf{A}_0 = \mathbf{B}_0 \times \mathbf{C}_0,$$

and

$$\text{Re}\,(\boldsymbol{\mathcal{A}}) = \mathbf{B}_0 \times \text{Re}\,(\boldsymbol{\mathcal{C}}) + \text{Re}\,(\boldsymbol{\mathcal{B}}) \times \mathbf{C}_0.$$

Since $\mathbf{A}_0$, $\mathbf{B}_0$, and $\mathbf{C}_0$ are all real, we can write

$$\mathbf{B}_0 \times \text{Re}\,(\boldsymbol{\mathcal{C}}e^{j\omega t}) = \text{Re}\,(\mathbf{B}_0 \times \boldsymbol{\mathcal{C}}e^{j\omega t}),$$

and

$$\mathbf{A}_0 + \text{Re}\,(\boldsymbol{\mathcal{A}}e^{j\omega t}) = \text{Re}\,(\mathbf{A}_0 + \boldsymbol{\mathcal{A}}e^{j\omega t})$$

and so on. Hence we can write

$$\text{Re}\,(\boldsymbol{\mathcal{A}}) = \text{Re}\,(\mathbf{B}_0 \times \boldsymbol{\mathcal{C}} + \boldsymbol{\mathcal{B}} \times \mathbf{C}_0).$$

Similarly, it may be shown that

$$\text{Im}\,(\boldsymbol{\mathcal{A}}) = \text{Im}\,(\mathbf{B}_0 \times \boldsymbol{\mathcal{C}} + \boldsymbol{\mathcal{B}} \times \mathbf{C}_0),$$

and therefore we finally have

$$\boldsymbol{\mathcal{A}} = \mathbf{B}_0 \times \boldsymbol{\mathcal{C}} + \boldsymbol{\mathcal{B}} \times \mathbf{C}_0.$$

The equation is thus split into two. The dc-equation is the same as the original and remains nonlinear, but the so-called ac-equation (although it still contains the dc-terms B_0 and C_0) is linear in the ac-components; i.e., it contains no powers or products of ac-components.

We now apply this approximation to the equation of motion of the electrons for the magneto-ionic medium. However, we assume that there is no net movement of the electrons, and so the dc-term in $\mathbf{v}$ is zero. Thus writing

$$\mathbf{B} = \mathbf{B}_0 + \text{Re}\,(\boldsymbol{\mathcal{B}}e^{j\omega t}), \qquad \mathbf{v} = \text{Re}\,(\hat{\mathbf{v}}e^{j\omega t}),$$

we have for the ac-equation,

$$j\omega m\hat{\mathbf{v}} = e\boldsymbol{\mathcal{E}} + e\hat{\mathbf{v}} \times \mathbf{B}_0.$$

Since we have assumed $\mathbf{v}$ to have sinusoidal time variation, the displacement $\mathbf{l}$, and therefore the polarization $\mathbf{P}$, will also have this form, and we can write

$$\boldsymbol{\mathcal{P}} = \text{N}e\hat{\mathbf{l}};$$

hence

$$\hat{\mathbf{v}} = j\omega\hat{\mathbf{l}} = \frac{j\omega}{Ne}\,\boldsymbol{\mathcal{P}},$$

and

$$j\omega\hat{\mathbf{v}} = -\frac{\omega^2}{Ne}\,\boldsymbol{\mathcal{P}}.$$

Substituting, we have

$$-\frac{m\omega^2}{Ne}\boldsymbol{\mathcal{P}} = e\boldsymbol{\mathcal{E}} + e\frac{j\omega}{Ne}\boldsymbol{\mathcal{P}} \times \mathbf{B}_0,$$

and writing this out in components, recalling that $\mathbf{B}_0$ is in the z-direction, we have

$$-\frac{m\omega^2}{Ne}\mathcal{P}_x = e\mathcal{E}_x + \frac{j\omega B_0}{N}\mathcal{P}_y,$$

$$-\frac{m\omega^2}{Ne}\mathcal{P}_y = e\mathcal{E}_y - \frac{j\omega B_0}{N}\mathcal{P}_x,$$

$$-\frac{m\omega^2}{Ne}\mathcal{P}_z = e\mathcal{E}_z.$$

In Section 8-2 we showed that the dielectric constant for the simple ionized medium is given by $\kappa = 1 - (Ne^2/\epsilon_0 m\omega^2)$. Hence for a given N there is a particular value of frequency, ω_N, known as the critical frequency which makes κ zero. This is given by $\omega_N^2 = Ne^2/\epsilon_0 m$. Thus we can write

$$-\frac{\omega^2}{\epsilon_0\omega_N^2}\mathcal{P}_x = \mathcal{E}_x + j\frac{\omega B_0}{Ne}\mathcal{P}_y = \mathcal{E}_x + j\frac{\omega^2}{\epsilon_0\omega_N^2}\frac{eB_0}{m\omega}\mathcal{P}_y.$$

But eB_0/m is the gyrofrequency (Section 11-1) ω_H. Defining the parameters

$$X = \frac{\omega_N^2}{\omega^2}, \qquad Y = \frac{\omega_H}{\omega},$$

we have

$$\mathcal{P}_x = -\epsilon_0 X\mathcal{E}_x - jY\mathcal{P}_y.$$

Similarly it may be shown that

$$\mathcal{P}_y = -\epsilon_0 X\mathcal{E}_y + jY\mathcal{P}_x, \qquad \mathcal{P}_z = -\epsilon_0 X\mathcal{E}_z.$$

Expressing $\mathcal{P}_x$, $\mathcal{P}_y$ in terms of $\mathcal{E}_x$, $\mathcal{E}_y$ gives

$$\mathcal{P}_x = -\frac{\epsilon_0 X}{1 - Y^2}(\mathcal{E}_x - jY\mathcal{E}_y), \qquad \mathcal{P}_y = -\frac{\epsilon_0 X}{1 - Y^2}(jY\mathcal{E}_x + \mathcal{E}_y).$$

Since $\boldsymbol{\mathcal{D}} = \epsilon_0\boldsymbol{\mathcal{E}} + \boldsymbol{\mathcal{P}}$, it follows that

$$\mathcal{D}_x = \epsilon_0(1 + C)\mathcal{E}_x - j\epsilon_0 CY\mathcal{E}_y,$$

where $C = -X/(1 - Y^2)$. Similarly

$$\mathcal{D}_y = j\epsilon_0 CY\mathcal{E}_x + \epsilon_0(1 + C)\mathcal{E}_y, \qquad \mathcal{D}_z = \epsilon_0(1 - X)\mathcal{E}_z.$$

In Chapter 8 we saw that we can take account of the effects of the medium either in terms of a permittivity and a permeability or in terms of a polarization and a magnetization. Here we have used the second method. The application of the electric field causes a displacement of the electrons, which in turn produces a dipole moment and hence a polarization which we calculated in components and then used the relation between $\mathfrak{D}$, $\mathcal{E}$, and $\mathcal{P}$ to obtain equations relating the components of $\mathfrak{D}$ and $\mathcal{E}$. We can still write $\mathfrak{D} = \epsilon\mathcal{E}$ symbolically, but now ϵ no longer has the simple interpretation that it had in Chapter 8. We must interpret it as the array of quantities

$$\epsilon_0 \begin{pmatrix} 1 + C & -jCY & 0 \\ jCY & 1 + C & 0 \\ 0 & 0 & 1 - X \end{pmatrix}.$$

The equation $\mathfrak{D} = \epsilon\mathcal{E}$ now represents the above set of three equations relating $\mathfrak{D}_x$, $\mathfrak{D}_y$, $\mathfrak{D}_z$ to $\mathcal{E}_x$, $\mathcal{E}_y$, $\mathcal{E}_z$. We can see the physical significance of this by supposing $\mathcal{E}_y = \mathcal{E}_z = 0$. We then have $\mathcal{E}$ in the x-direction. The above relationships show that $\mathfrak{D}$ has a y-component and hence is not in the x-direction. Thus $\mathfrak{D}$ and $\mathcal{E}$ are in different directions. The permittivity is now a more complicated type of quantity known as a *tensor*. It is not too difficult to get some idea of what a tensor is. If we take a vector quantity such as $\mathcal{E}$ and operate on it with (i.e., multiply it by) a scalar ϵ, we get a new vector $\mathfrak{D}$ which has a different magnitude but is in the same direction. But if we operate on it with a tensor ϵ, then the resulting $\mathfrak{D}$-vector is different in both magnitude and direction. We can see physically how this happens in the present case. When the electric field is applied, the electrons begin to move, but are then deflected by the magnetic field. Thus the resulting displacement is not in the same direction as the field. Consequently the polarization and the electric field are in different directions, and, when they are added vectorially, we get a flux density which is in a different direction from the electric field. We noted in Section 3–5 that media in which **D** and **E** are in the same direction are called isotropic. For the magneto-ionic medium, this does not hold; here we have an example of an anisotropic medium.

12–3 Waves in magneto-ionic media

Like all other fields, those in a magneto-ionic medium must obey the equations developed in Chapter 5. The two Maxwell equations and the three constitutive relations are sufficient to solve the problem, which however, is now made more difficult by the complicated nature of the constitutive relations derived above. We can no longer keep our work in terms of vectors since this would necessitate a knowledge of how to manipulate the tensor which would automatically be introduced. We therefore work in terms of Cartesian coordinates. This makes the procedure more cumbersome and lengthy, and we must proceed systematically to get our results.

We proceed along lines very similar to those followed in Section 9–3, and assume that all quantities have sinusoidal time variation and that they are functions of z only, so that $\partial/\partial x = \partial/\partial y = 0$. In addition, we suppose that there are no losses and that waves traveling in the z-direction can, in fact, exist so that we can assume a z-variation $e^{-j\beta z}$. Maxwell's equations may be written

$$\text{curl } \mathbf{H} = \partial \mathbf{D}/\partial t, \qquad \text{curl } \mathbf{E} = -\partial \mathbf{B}/\partial t.$$

With the above assumptions, we have

$$\text{curl } \boldsymbol{\mathcal{H}} = j\omega \boldsymbol{\mathcal{D}}, \qquad \text{curl } \boldsymbol{\mathcal{E}} = -j\omega \boldsymbol{\mathcal{B}},$$

and in rectangular components,

$$j\beta \mathcal{H}_y = j\omega \mathcal{D}_x, \qquad -j\beta \mathcal{E}_y = j\omega \mathcal{B}_x,$$

$$-j\beta \mathcal{H}_x = j\omega \mathcal{D}_y, \qquad j\beta \mathcal{E}_x = j\omega \mathcal{B}_y.$$

Using the constitutive relations derived in the last section and assuming that the medium is nonmagnetic, we obtain

$$j\beta \mathcal{H}_y = j\omega\epsilon_0[(1 + C)\mathcal{E}_x - jYC\mathcal{E}_y], \qquad -j\beta \mathcal{E}_y = j\omega\mu_0 \mathcal{H}_x,$$

$$-j\beta \mathcal{H}_x = j\omega\epsilon_0[jYC\mathcal{E}_x + (1 + C)\mathcal{E}_y], \qquad j\beta \mathcal{E}_x = j\omega\mu_0 \mathcal{H}_y.$$

Eliminating $\mathcal{H}_x$ and $\mathcal{H}_y$ from these four equations yields

$$\left[\frac{\beta^2}{j\omega\mu_0} + j\omega\epsilon_0(1 + C)\right] \mathcal{E}_x = -\omega\epsilon_0 YC\mathcal{E}_y,$$

$$\left[\frac{\beta^2}{j\omega\mu_0} + j\omega\epsilon_0(1 + C)\right] \mathcal{E}_y = \omega\epsilon_0 YC\mathcal{E}_x,$$

from which it follows that $\mathcal{E}_x/\mathcal{E}_y = -\mathcal{E}_y/\mathcal{E}_x$. Hence

$$\mathcal{E}_x = \pm j\mathcal{E}_y.$$

The time-varying fields are then given by

$$E_x = \text{Re}\,(\mathcal{E}_x e^{j\omega t}) = E_0 \cos\,(\omega t - \beta z + \alpha),$$

where E_0 is the magnitude of E_x, and α is its phase. We can choose $t = 0$ so that $\alpha = 0$, and hence $E_x = E_0 \cos\,(\omega t - \beta z)$, and for $\mathcal{E}_x = j\mathcal{E}_y$,

$$E_y = \text{Re}\,(\mathcal{E}_y e^{j\omega t}) = \text{Re}\,(-j\mathcal{E}_x e^{j\omega t}) = E_0 \sin\,(\omega t - \beta z).$$

The magnitude of the resultant is thus

$$\sqrt{(|E_x|^2 + |E_y|^2)} = E_0,$$

and the angle θ which it makes with the x-axis is

$$\theta = \tan^{-1}\frac{E_y}{E_x} = (\omega t - \beta z).$$

This vector has a constant magnitude, and for a given value of z (i.e., at a fixed point in space), the angle increases uniformly with time. The tip of the vector therefore traces out a circle in the xy-plane, and the wave is circularly polarized. The direction of rotation is related to the direction of propagation by the right-hand rule, and we shall refer to this as right-handed polarization.* In the same way we are able to show that $\mathcal{E}_x = -j\mathcal{E}_y$ is a left-handed circularly polarized wave.

The result is surprising at first sight. We seem to have proved that the only waves which can exist in the magneto-ionic medium are those with circular polarization. This does not seem to be right. After all, we could have a linearly polarized wave incident upon such a medium (for example, the ionization might be contained in a glass tube of negligible wall thickness and the wave be incident on it from the surrounding air, where we know linear polarization to be possible). When we arrive at such a situation, we should carefully review our assumptions and see what restrictions these impose on our result. In this case the relevant assumption turns out to be that of $e^{-j\beta z}$ for the z-variation. If all components of the field, and in particular E_x and E_y, vary in exactly the same way with z, then they bear the same relation to each other for all values of z, that is, the polarization of the wave is independent of z. We saw in the case of the waveguide that the polarization varied across the cross section of the guide, i.e., it varied with x and y. There is no fundamental reason why it should not vary with z also, and in the general case it does. Therefore what we have derived is the polarization which a wave must have if it is to propagate without change of polarization. A wave which propagates without change of polarization is known as a *characteristic wave*.

We can now begin to answer the above question. We can regard a linear polarization as composed of a superposition of a left- and a right-handed circular polarization of equal magnitude. (We are more accustomed to analyzing a circular polarization in terms of two linear polarizations at right angles, but the opposite process needed here is obviously equally valid.) The linearly polarized wave is split into two components as soon as it enters the medium. If we superpose these after they have propagated some distance into the medium, they will still produce linear polarization. But if our reasoning is correct, this polarization must be different from that of the incident wave. Since we see that the polarization is always linear, it is only the direction of the polarization that can vary. This arises as follows. Using formulas obtained in the derivation above, we have

$$\left[\frac{\beta^2}{j\omega\mu_0} + j\omega\epsilon_0(1 + C)\right]\mathcal{E}_x = -\omega\epsilon_0 YC\mathcal{E}_y.$$

* This convention is not universal. Most engineering authors follow it, but workers in optics often define "right-handed" to mean clockwise as viewed by the observer, which is opposite to our definition.

For the right-handed polarization,

$$\mathcal{E}_x = j\mathcal{E}_y,$$

and thus

$$\frac{\beta^2}{j\omega\mu_0} + j\omega\epsilon_0(1 + C) = j\omega\epsilon_0 YC.$$

Solving for β and substituting for C, we have

$$\beta^2 = \omega^2\epsilon_0\mu_0\left(1 - \frac{X}{1 + Y}\right).$$

We might pause and note at this point that we have eliminated *all* of our variables, i.e., we have eliminated six unknowns with only six equations. Now we normally need $n + 1$ equations to eliminate n unknowns. However, since the original equations are homogeneous, we know that if they are to have a nontrivial solution there must be a relationship between the coefficients, namely, the determinant of the coefficients must be zero. We have, in effect, obtained the determinantal equation above. Returning now to this equation, we find that the phase constant β is given by

$$\beta^2 = \omega^2\epsilon_0\mu_0\left(1 - \frac{X}{1 + Y}\right).$$

Now

$$\frac{\beta^2}{\omega^2\epsilon_0\mu_0} = \frac{c^2}{v^2} = n^2,$$

where c^2 is the velocity of light in vacuum, v^2 is the phase velocity, and n is the refractive index. Hence

$$n^2 = 1 - \frac{X}{1 + Y}.$$

Similarly, it may be shown that for the left-handed wave,

$$n^2 = 1 - \frac{X}{1 - Y}.$$

Thus it is seen that the two characteristic waves have different refractive indexes and so different phase velocities. If our linearly polarized wave is incident normally upon the magneto-ionic medium, then it is split into two circularly polarized waves traveling at different velocities. Hence their relative phases change with z (Fig. 12-1). The incident wave is split into the two circular components at the interface. These two vectors rotate at the same rate, and at the point O, the resultant is a linear oscillation in the direction shown. We next suppose that the upper component has the slower velocity. It will accordingly take longer to reach the point X than the lower one. Since the two rotate

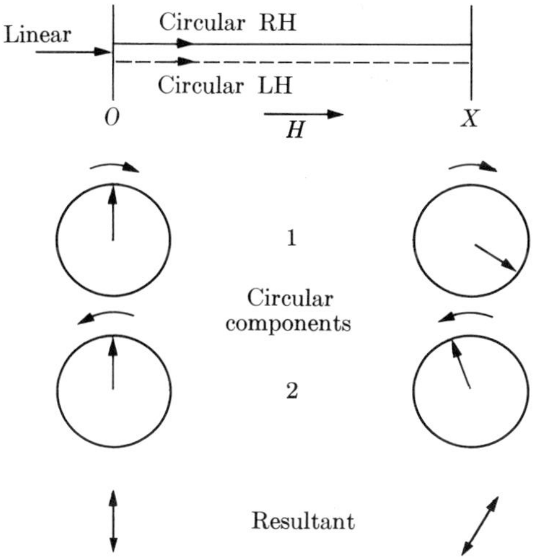

FIG. 12–1. The rotation of the direction of polarization.

at the same speed, the upper one will sweep through a larger angle in the time it takes to get to X. This situation is shown in the figure, which represents a snapshot of the situation at one instant of time and shows the direction of the component vectors at O and X at that instant. If we let time vary, they all rotate at the same speed. The two components at O and X combine to give linear oscillations, but the direction of the vibrations is seen from the figure to be different. The angle through which this direction is turned increases as we progress in the z-direction, and so the direction of polarization rotates. This effect is well known in optics. Substances exhibiting the effect are said to be optically active and to rotate the plane of polarization of polarized light.

If the wave is not incident normally on the surface, then owing to the difference in the refractive indices the two components are refracted by different amounts. In this case the splitting of the incident ray into two components is much more obvious, since the two are now traveling in different directions. This effect is also well known in optics. It is exhibited by crystals, and is known as *double refraction.*

12–4 Ferrites and their properties

The ferrites are a group of chemical compounds to which lodestone belongs. It will be recalled that much of the early work on magnetism followed the discovery of this material and its properties. Chemically we may write lodestone, the magnetic oxide of iron, as $FeO \cdot Fe_2O_3$. The general formula for the ferrites is $MO \cdot Fe_2O_3$, where M is a metal such as manganese, zinc, or nickel. The ferrites have some special and very important properties from the viewpoint of electromagnetism. They have a relatively high resistivity in the range 10^2–10^6

ohm-cm as compared to 10^{-5} for iron and 10^{16} for mica. They have high permeabilities of the order of 1000, but this is not so high as that for the ferromagnetic materials, whose permeabilities are of the order of 10,000. They also have high dielectric constants in the range 2–25. The ferrites have the unique combination of high resistivity and high permeability, in contrast to the dielectrics, which have permeabilities near that of vacuum, and the ferromagnetics which, although exhibiting a much higher permeability, also exhibit large losses. The high resistivity makes it possible for the ferrites to support electric fields, and wave motion may occur in them. We must first, however, deduce the constitutive relations for the ferrites, and we shall see that in the presence of a magnetic polarizing field, these compounds exhibit a magnetic anisotropy very similar to the electric anisotropy of the magneto-ionic medium which we have just studied.

A complete treatment of the magnetism exhibited by ferrites requires the use of quantum theory. However, we can obtain sufficient information for our purposes by using a classical model. Experimental atomic physics has shown that the electron possesses both a magnetic dipole moment and an angular momentum. If we imagine the electron as a sphere with a uniform charge distribution, rotating about an axis through its center, then owing to the rotating mass it would possess angular momentum, and the rotating charge would constitute a current loop and would be equivalent to a magnetic dipole. Furthermore, since the charge is negative, if we represent the angular momentum by a vector whose direction is determined by the right-hand rule, the angular momentum vector will be in the opposite direction to the magnetic dipole moment (the two are said to be antiparallel). This, in fact, agrees with experimental results, and since these can be explained by our rotating-sphere model, the electron is said to possess a spin. The arrangement of the electrons in the molecule and of the molecules in the solid structure can have the result that all spins cancel, in which case the material does not exhibit magnetic properties, or there can be a resultant spin as there is in the case of the ferrites.

If $\mathbf{m}$ is the magnetic dipole moment of the electron, and $\mathbf{J}$ is its angular momentum, then experiment shows that

$$\mathbf{m} = \gamma \mathbf{J},$$

where γ is a constant known as the *gyromagnetic ratio*. We suppose that we have a ferrite with a polarizing magnetic field applied to it so that all the spins are approximately lined up with the field. We saw in Chapter 4 that the couple on a dipole in a magnetic field is given by

$$\mathbf{T} = \mathbf{m} \times \mathbf{B}.$$

But the couple is equal to the rate of change of angular momentum, and thus

$$\mathbf{T} = \frac{d\mathbf{J}}{dt} = \frac{1}{\gamma}\frac{d\mathbf{m}}{dt} = \mathbf{m} \times \mathbf{B}.$$

The magnetization **M** (Chapter 8) will be proportional to the magnetic moment **m**, and hence

$$\frac{d\mathbf{M}}{dt} = \gamma \mathbf{M} \times \mathbf{B}.$$

The field **B** acting on the dipoles will be the field within the medium. We saw in Chapter 8 that we can represent the effect of the medium either by a permeability or in terms of its magnetization. In this case, it is clearly more appropriate to use magnetization. Since we supposed that all spins are approximately lined up with the applied field, to a good approximation we can write (as in Chapter 8)

$$\mathbf{B} = \mu_0(\mathbf{H} + \mathbf{M}),$$

where **H** is the applied polarizing field. Hence

$$\frac{d\mathbf{M}}{dt} = \gamma\mu_0 \mathbf{M} \times (\mathbf{H} + \mathbf{M}),$$

and since $\mathbf{M} \times \mathbf{M} = 0$, we have

$$\frac{d\mathbf{M}}{dt} = \gamma\mu_0 \mathbf{M} \times \mathbf{H}.$$

It is interesting to consider the motion of the spin. Any small misalignment between the magnetic moment and the polarizing field will result in a couple. Owing to the angular momentum, this will not cause the spin to move into alignment with the field, but will cause it to precess about the field. We can picture this in terms of the gyroscope. We know that if the axis of a spinning gyroscope is in a slightly off-vertical position, it sweeps out a cone about the vertical. This motion, called *precession*, is a result of the couple produced by the mechanical mass of the gyroscope interacting with the gravitational field (which is vertical). The precession is about the direction of the field responsible for the couple. In our case, the couple arises from the interaction of the magnetic dipole with the magnetic field, and thus is magnetic in origin rather than mechanical. However, the result is a mechanical couple, which interacts with the angular momentum and produces precession—in this case about the direction of the magnetic field. We know from mechanics that the velocity of precession is given by $T/(J \sin\theta)$, where θ is the angle between the angular momentum vector **J** and the direction about which it is precessing. Hence the precessional velocity is given here by

$$\omega_0 = \frac{T}{J\sin\theta} = \frac{\mu_0 m H \sin\theta}{(m/\gamma)\sin\theta} = \mu_0\gamma H.$$

This is known as *Larmor precession*, and the frequency is called the *gyromagnetic resonance frequency*. It is the counterpart of the gyrofrequency in the case of the magneto-ionic medium.

Returning now to the equation

$$\frac{d\mathbf{M}}{dt} = \gamma\mu_0\mathbf{M} \times \mathbf{H},$$

we note that this is a nonlinear equation, since it contains the product $\mathbf{M} \times \mathbf{H}$, which is just the type of term we can deal with by using the small-signal approximation developed in the last section. The **H**-vector has a steady component (the polarizing field) which will produce a steady component of **M** by virtue of the tendency to align the dipoles. Thus supposing that both **H** and **M** consist of a steady component and an ac-component, and applying the small-signal approximation to the equation for the ferrite, we have

$$\frac{d}{dt}(\mathbf{M}_0 + \mathrm{Re}\, \mathfrak{M}e^{j\omega t}) = \gamma\mu_0(\mathbf{M}_0 + \mathrm{Re}\, \mathfrak{M}e^{j\omega t}) \times (\mathbf{H}_0 + \mathrm{Re}\, \mathfrak{H}e^{j\omega t}).$$

Hence

$$0 = \mathbf{M}_0 \times \mathbf{H}_0, \qquad \text{and} \qquad j\omega\mathfrak{M} = \gamma\mu_0(\mathbf{M}_0 \times \mathfrak{H} + \mathfrak{M} \times \mathbf{H}_0).$$

Note that the magnetization produced by the polarizing field is in the same direction as the field, as was explained above, and hence the cross product must be zero. This checks the first of the above equations.

Now let us choose a set of axes with the z-axis in the direction of $\mathbf{H}_0$ (and therefore also of $\mathbf{M}_0$). Writing the above vector equation in components, we obtain

$$\begin{aligned} j\omega\mathfrak{M}_x &= \gamma\mu_0(\mathfrak{M}_y H_0 - \mathfrak{H}_y M_0), \\ j\omega\mathfrak{M}_y &= \gamma\mu_0(-\mathfrak{M}_x H_0 + \mathfrak{H}_x M_0), \\ j\omega\mathfrak{M}_z &= 0. \end{aligned}$$

The first two are a pair of simultaneous equations involving $\mathfrak{H}_x$, $\mathfrak{H}_y$, $\mathfrak{M}_x$, and $\mathfrak{M}_y$. We can eliminate any one of these variables from the equations by suitable manipulation and produce a single equation involving the other three. Eliminating $\mathfrak{M}_y$ gives

$$\mathfrak{M}_x(\gamma^2\mu_0^2H_0^2 - \omega^2) = \gamma^2\mu_0^2M_0H_0\mathfrak{H}_x - j\omega\gamma\mu_0M_0\mathfrak{H}_y.$$

We see that $\gamma\mu_0H_0$ is just ω_0, the Larmor frequency, and we define a new quantity $\omega_M = \gamma\mu_0M_0$. Hence

$$\mathfrak{M}_x = \frac{\omega_0\omega_M}{\omega_0^2 - \omega^2}\mathfrak{H}_x - j\frac{\omega\omega_M}{\omega_0^2 - \omega^2}\mathfrak{H}_y.$$

But the flux density $\mathfrak{B}$ is given by $\mathfrak{B} = \mu_0(\mathfrak{H} + \mathfrak{M})$, and thus

$$\mathfrak{B}_x = \mu_0\left(1 + \frac{\omega_0\omega_M}{\omega_0^2 - \omega^2}\right)\mathfrak{H}_x - j\mu_0\frac{\omega\omega_M}{\omega_0^2 - \omega^2}\mathfrak{H}_y.$$

Finally defining the two new parameters

$$\mu = \mu_0\left(1 + \frac{\omega_0\omega_M}{\omega_0^2 - \omega^2}\right) \qquad \text{and} \qquad \kappa = \mu_0\,\frac{\omega\omega_M}{\omega_0^2 - \omega^2},$$

we have

$$\mathcal{B}_x = \mu\mathcal{H}_x - j\kappa\mathcal{H}_y.$$

Similarly it may be shown that

$$\mathcal{B}_y = j\kappa\mathcal{H}_x + \mu\mathcal{H}_y.$$

Since $\mathcal{M}_z = 0$,

$$\mathcal{B}_z = \mu_0\mathcal{H}_z.$$

Hence we can write

$$\boldsymbol{\mathcal{B}} = \mu\boldsymbol{\mathcal{H}},$$

where μ is the permeability tensor

$$\mu = \begin{pmatrix} \mu & -j\kappa & 0 \\ j\kappa & \mu & 0 \\ 0 & 0 & \mu_0 \end{pmatrix}.$$

We now have the constitutive relations for the ferrite with a polarizing magnetic field.

12–5 Waves in ferrites

In this section we shall investigate waves in polarized ferrite. We shall consider propagation parallel to the polarizing field (the same case as we treated for the magneto-ionic medium). We assume that there are no losses and that the medium is infinite, so that $\partial/\partial x = \partial/\partial y = 0$. Maxwell's equations can now be written

$$\text{curl}\,\mathbf{H} = \epsilon\frac{\partial\mathbf{E}}{\partial t}, \qquad \text{curl}\,\mathbf{E} = -\frac{\partial\mathbf{B}}{\partial t}.$$

Assuming sinusoidal time variation

$$\text{curl}\,\boldsymbol{\mathcal{H}} = j\omega\epsilon\boldsymbol{\mathcal{E}}, \qquad -\text{curl}\,\boldsymbol{\mathcal{E}} = j\omega\boldsymbol{\mathcal{B}},$$

and supposing a z-variation $e^{-j\beta z}$, we have, in Cartesian coordinates,

$$\begin{aligned} j\beta\mathcal{H}_y &= j\omega\epsilon\mathcal{E}_x, & -j\beta\mathcal{E}_y &= j\omega\mathcal{B}_x, \\ -j\beta\mathcal{H}_x &= j\omega\epsilon\mathcal{E}_y, & j\beta\mathcal{E}_x &= j\omega\mathcal{B}_y, \\ 0 &= j\omega\epsilon\mathcal{E}_z, & 0 &= j\omega\mathcal{B}_z. \end{aligned}$$

From the constitutive relations, we have

$$\mathcal{B}_x = \mu\mathcal{H}_x - j\kappa\mathcal{H}_y,$$
$$\mathcal{B}_y = j\kappa\mathcal{H}_x + \mu\mathcal{H}_y,$$

and substituting these into Maxwell's equations, we obtain

$$j\beta\mathcal{H}_y = j\omega\epsilon\mathcal{E}_x, \qquad -j\beta\mathcal{E}_y = j\omega(\mu\mathcal{H}_x - j\kappa\mathcal{H}_y),$$
$$-j\beta\mathcal{H}_x = j\omega\epsilon\mathcal{E}_y, \qquad j\beta\mathcal{E}_x = j\omega(j\kappa\mathcal{H}_x + \mu\mathcal{H}_y).$$

Eliminating $\mathcal{E}_x$ from these equations gives

$$(\beta^2 - \omega^2\epsilon\mu)\mathcal{H}_y = j\kappa\omega^2\epsilon\mathcal{H}_x,$$

and eliminating $\mathcal{E}_y$ gives

$$(\beta^2 - \omega^2\epsilon\mu)\mathcal{H}_x = -j\kappa\omega^2\epsilon\mathcal{H}_y,$$

from which it follows that $\mathcal{H}_x^2 = -\mathcal{H}_y^2$ or $\mathcal{H}_x = \pm j\mathcal{H}_y$. Since we supposed that all quantities vary as $e^{-j\beta z}$ with z, then the waves which result from the analysis will be characteristic waves, which are two in number and have right- and left-handed circular polarization. To obtain β we use the result $\mathcal{H}_x = \pm j\mathcal{H}_y$ giving

$$(\beta^2 - \omega^2\epsilon\mu) = \pm\kappa\omega^2\epsilon,$$
$$\beta = \pm\omega\sqrt{\epsilon(\mu \pm \kappa)},$$

and the phase velocity

$$v_p = \frac{\omega}{\beta} = \pm\frac{1}{\sqrt{\epsilon(\mu \pm \kappa)}},$$

and again the two characteristic waves travel with different velocities. The polarized ferrite will also have the property of rotating the direction of polarization of a linearly polarized wave.

12-6 The isolator and the gyrator

The rotation of the direction of polarization has been used to produce microwave devices such as the isolator and the gyrator. These have the property that they are not reciprocal devices. Let us consider the phase velocities a little more closely. If we put $\mathcal{H}_x = j\mathcal{H}_y$, the direction of rotation will be right-handed relative to the z-axis or to the direction of the polarizing field. The phase constant is then given by $\beta = \pm\omega\sqrt{\epsilon(\mu - \kappa)}$ and the phase velocity by $v_p = \pm 1/\sqrt{\epsilon(\mu - \kappa)}$, which is the same for propagation in the $+z$- and the $-z$-directions. However, if we relate the polarization to the direction of propagation, as we did in Section 12-3, then the wave traveling in the $-z$-direction has left-handed polarization, so the right-handed wave travels in the $+z$-direction with the same velocity as the left-handed wave does in the $-z$-direction. We can see the effect that this produces by reference to Fig. 12-2, which is an elaboration of Fig. 12-1. In the upper diagram we have a wave propagating from left to right. This is split into two circular components in the ferrite, and, if we suppose component 1 to have the slower velocity, we have the same effect as described for the magneto-ionic medium. We assume that this wave is reflected back through the ferrite. Component 1 still has the smaller velocity since we just saw that the phase velocity is the same in the $+z$- and $-z$-directions for the same sense of rotation relative to the polarizing field. The field vector for this component therefore rotates through the larger angle in traveling back to O, and the effect is shown in the lower diagram. Thus the polarization is not restored to its original direction when the wave travels back through the ferrite. The angle through which the direction of polarization is rotated is doubled, not canceled. Therefore the device does not obey the principle of reciprocity.

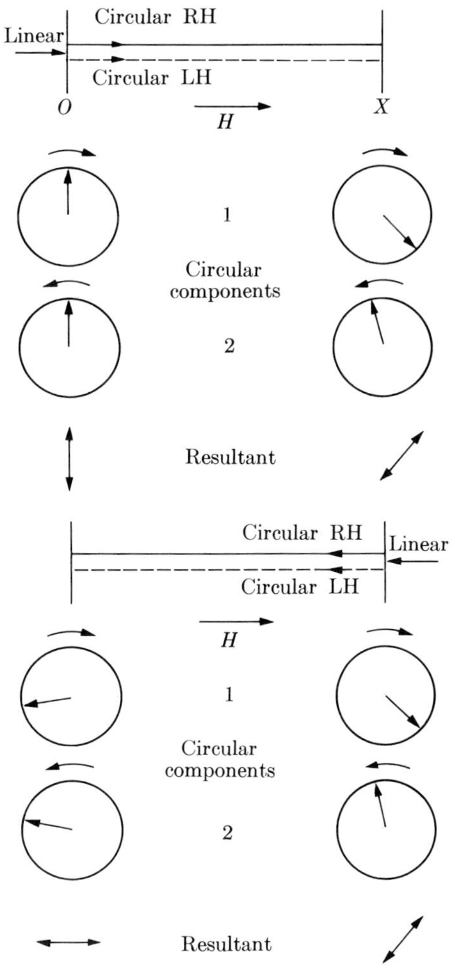

FIG. 12-2. A nonreciprocal device.

The analysis in the last section is for an infinite medium which cannot, of course, be used in practical engineering. It may be shown that for the case of a circular ferrite rod inside a circular waveguide with a polarizing magnetic field along the axis of the guide, the characteristic waves are also circularly polarized just as they are for the infinite medium, and so the same effects will be exhibited. The action of the isolator can be explained by reference to Fig. 12-3. A rectangular waveguide is connected to a circular guide by means of a section whose cross section varies slowly within one wavelength so that a good match is obtained between the guides. The circular guide contains a ferrite rod tapered at the ends to minimize reflections, and a permanent magnet provides the polarizing field. The circular guide then connects with a second rectangular guide with the same cross section as the first, but with its sides rotated 45° relative to this guide. This is shown in the sections in Fig. 12-3. We suppose that the left-hand

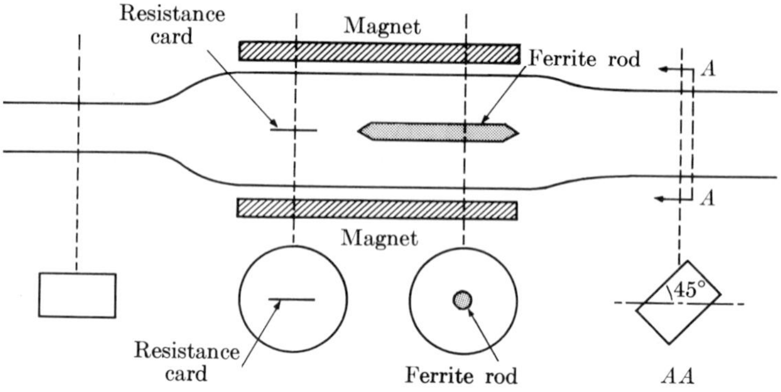

FIG. 12-3. The isolator.

guide is excited in the TE_{10}-mode, which is linearly polarized, with the electric vector vertical. The direction of polarization will be rotated by the ferrite as described above, and we suppose that the direction of the polarizing field and the length of the rod are so chosen that this rotation is 45° in such a direction that the field is now correctly orientated to propagate as a TE_{10}-mode in the right-hand guide. Let us next consider a TE_{10}-mode excited in the right-hand guide. In traveling through the ferrite its direction of polarization will be rotated 45°, so that it will arrive at the left-hand guide with its electric vector horizontal. This would then excite a TE_{01}-mode in the left-hand guide. If we design the guides so that only the dominant mode propagates, then this mode will be cut off and no propagation is possible. The energy in the wave would be reflected, but this is prevented in practice by the use of a horizontal resistance card placed as shown in the figure. The wave arriving from the right has a horizontal electric vector and thus will produce currents in the resistance which dissipate its energy. The TE_{10}-mode from the left, however, has a vertical electric vector, and hence no currents are induced in the card and the wave passes unattenuated. Thus a wave can propagate from left to right through the device but not from right to left. This device is used to isolate a source from its load, for example, so that reflections from the load due to mismatch cannot feed back into the generator. In practice it is often convenient for the input and output guides to be parallel rather than at an angle of 45°. This can be accomplished by means of a 45° twist in the guide, as illustrated in Fig. 12-4. The field will "follow the guide," i.e., the electric field lines will stay parallel to the short side of the guide (assuming a TE_{10}-mode), provided the amount of twist in one wavelength is small.

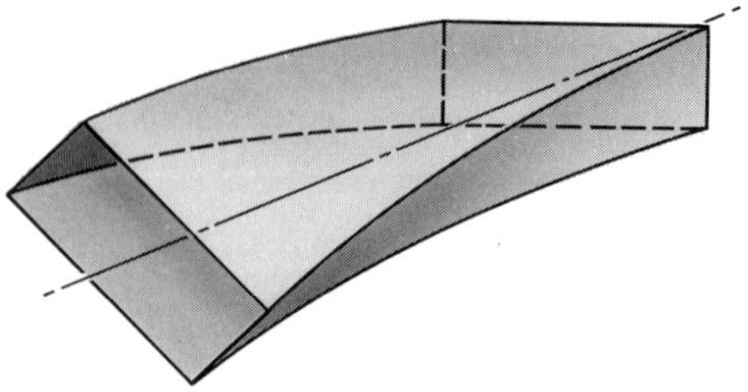

FIG. 12-4. A 45° twist section.

Another device based on these same principles employs rotations of 90°. Referring positive and negative rotations to the direction of the z-axis, we find that a wave traveling in the $+z$-direction receives $+90°$ rotation by the ferrite and $-90°$ by the twist in the guide, so that the overall rotation is zero. A wave traveling in the $-z$-direction is rotated $+90°$ by the twist and $+90°$ by the ferrite, the overall rotation being 180° in this case, which is equivalent to a *time* phase shift of 180°. Thus the device has a zero phase shift in one direction and a phase shift of 180° in the other. We may compare this with the gyrator which in circuit theory is defined as a four-terminal network described by the following equations,

$$V_1 = -sI_2, \qquad V_2 = sI_1.$$

If the gyrator is driven at port 1 and has a resistive load so that $V_2/I_2 = R$, then substitution shows that $V_1 = -(s/R)V_2$. If it is driven from port 2, then $V_1/I_1 = R$ and $V_2 = (s/R)V_1$. Thus V_1 and V_2 are in phase if the signal passes through the gyrator in one direction, and they have a phase difference of 180° if it passes through in the opposite direction. The device described in the previous paragraph is closely related to the gyrator. In fact, it is sometimes referred to as the microwave gyrator.

Problems

12–1 Use the small-signal approximation to linearize the equation $y = x^2$.

12–2 Repeat Problem 12–1 for the equation $\mathbf{A} = \mathbf{B} \cdot \mathbf{C}$.

12–3 Repeat Problem 12–1 for the equation $\mathbf{A} = \mathbf{B} \cdot (\partial \mathbf{C}/\partial t)$.

12–4 Using the equations developed in Section 12–2, obtain a set of equations expressing $\mathcal{P}_x$, $\mathcal{P}_y$, $\mathcal{P}_z$ in terms of $\mathcal{E}_x$, $\mathcal{E}_y$, $\mathcal{E}_z$. Discuss the equation $\mathcal{P} = \epsilon_0 \chi \boldsymbol{\mathcal{E}}$ in this case.

12–5 A linearly polarized wave with $\boldsymbol{\mathcal{E}}$ in the x-direction and traveling in the z-direction is incident on a ferrite medium at $z = 0$. The steady magnetic field is in the z-direction. Calculate the angle between the $\boldsymbol{\mathcal{E}}$-vector and the x-direction at $z = L$ assuming $\mu \gg \kappa$.

12–6 A radar transmitter in a satellite sends a pulse vertically downward to the earth over the magnetic North Pole. The satellite is above the ionized regions of the atmosphere and the frequency is such that the wave can pass through the ionized layer. What would you expect to receive at the receiver on the ground, and how do you explain this result?

12–7 Calculate the gyrofrequency for an electron in the magnetic field of the earth which has a value of 35 amp/m. Calculate the electron densities required to produce a zero refractive index for the case of propagation parallel to the magnetic field in the earth's ionosphere at a frequency of 3 Mc/sec [$\epsilon_0 = (1/36\pi) \times 10^{-9}$ f/m, $\mu_0 = 4\pi \times 10^{-7}$ h/m, $e = 1.6 \times 10^{-19}$ coul, $m = 9.1 \times 10^{-31}$ kg].

12–8 Derive the formula for the precessional velocity. (Allow the torque to act for a time δt. This produces an increment of angular momentum which must be added vectorially to the spin momentum. You will find that this turns the spin-momentum vector through an angle $\delta\phi$, and the precessional velocity is accordingly $\delta\phi/\delta t$.)

12–9 Derive the magnetic susceptibility tensor for the ferrite in the case discussed in Section 12–4.

12–10 The problem treated in Section 12–5 is that of a wave traveling in a ferrite in the presence of a steady magnetic field parallel to the direction of propagation. The result shows that there are two characteristic waves with righ,- and left-handed circular polarization traveling at different phase velocities. Without resorting to the use of any coordinate axes, describe which polarization travels faster, for propagation in the same direction as the field and in the opposite direction. (The use of coordinates is permitted in the course of any deductions you may wish to make, but they must not appear in your final answer.)

12–11 The isolator produces a rotation of 90° of the field for a wave propagating in one direction and 0° for a wave propagating in the opposite direction. Devise a means of producing a rotation of 180° in one direction and 90° in the other.

12–12 Calculate the length of a ferrite rod required to produce an isolator at 10,000 Mc/sec, given that $f_M\,(=\omega_M/2\pi)$ for the ferrite material is 1000 Mc/sec and the gyromagnetic resonance frequency is 3000 Mc/sec.

CHAPTER 13

Electron Beams

13–1 Introduction

In Chapter 11 we considered the motion of charged particles in quasistatic fields. In the first part of the chapter we neglected the fields due to the particles themselves and later included the effects of space charge in our calculations. In the last chapter we considered the case of rapidly varying fields, but in our consideration of the magneto-ionic medium we neglected the effect of the fields of the particles. In this chapter we consider the case where the fields are rapidly varying, and the effects of space charge fields cannot be neglected. In solving the equations our approach will follow closely that of our previous work. We shall look for propagating waves, and to simplify the problem we shall assume that the medium is infinite and a uniform plane solution exists, so that we can put $\partial/\partial x = \partial/\partial y = 0$. Other simplifying assumptions will be introduced as we progress.

13–2 The equations for the beam

Our medium consists of a beam of electrons which we suppose to be moving in the z-direction, and the cross section of the beam is assumed to be infinite (i.e., in the x- and y-directions). The presence of the charges will give rise to an electric field, and their motion constitutes a current which will give rise to a magnetic field. These fields must obey Maxwell's equations, and thus we have

$$\text{curl}\,\mathbf{E} = -\mu_0 \frac{\partial \mathbf{H}}{\partial t}, \qquad \text{curl}\,\mathbf{H} = \epsilon_0 \frac{\partial \mathbf{E}}{\partial t} + \mathbf{J},$$

$$\text{div}\,\mathbf{E} = \frac{q_v}{\epsilon_0}, \qquad \text{div}\,\mathbf{H} = 0.$$

Taking the divergence of the first equation gives

$$\mu_0 \frac{\partial}{\partial t} \text{div}\,\mathbf{H} = 0,$$

and by the same argument as used in Section 5–2 we have div $\mathbf{H} = 0$. The fourth equation is consequently contained in the first. Taking the divergence of the second curl equation, we obtain

$$\epsilon_0 \frac{\partial}{\partial t} \text{div}\,\mathbf{E} + \text{div}\,\mathbf{J} = 0,$$

and substituting for div **E**, we have

$$\frac{\partial q_v}{\partial t} + \operatorname{div} \mathbf{J} = 0.$$

This is the equation of continuity obtained in Chapter 1. Thus far we have three independent equations, the two curl equations and the equation of continuity. (We could, of course, have kept the original three Maxwell equations rather than combining them so as to produce the equation of continuity as our third equation. However, this procedure proves to be more convenient.) Our three equations contain four unknowns, namely, **E**, **H**, **J**, and q_v. In Section 11-5, we derived the relation between **J** and q_v:

$$\mathbf{J} = q_v \mathbf{v}.$$

However, this introduces another unknown variable **v**, and so we still need another equation, which is provided by the equation of motion of the electron derived in Chapter 11:

$$m \frac{d\mathbf{v}}{dt} = e(\mathbf{E} + \mathbf{v} \times \mathbf{B}).$$

We now have five equations to solve for the five unknown quantities **E**, **H**, **J**, q_v, **v**. The next simplification which we introduce is to impose a steady magnetic focusing field in the z-direction. The electrons are now free to move in this direction without any restraining force due to the field. However, any motion perpendicular to the z-direction will result in a circular motion about the field. As we saw in Section 11-1, the radius of the circle is given by $mv_\perp/eB$. If we make the **B**-field very large, then the radius of the circle can be made negligibly small, and we may suppose that the motion of the electrons is confined to the z-direction.

Putting into our original five equations the assumptions we have made so far, namely, that $\partial/\partial x = \partial/\partial y = 0$ and that **v**, and therefore **J**, have only z-components, we have

$$\frac{-\partial E_y}{\partial z} = -\mu_0 \frac{\partial H_x}{\partial t}, \qquad -\frac{\partial H_y}{\partial z} = \epsilon_0 \frac{\partial E_x}{\partial t},$$

$$\frac{\partial E_x}{\partial z} = -\mu_0 \frac{\partial H_y}{\partial t}, \qquad \frac{\partial H_x}{\partial z} = \epsilon_0 \frac{\partial E_y}{\partial t},$$

$$0 = -\mu_0 \frac{\partial H_z}{\partial t}, \qquad 0 = \epsilon_0 \frac{\partial E_z}{\partial t} + J,$$

$$\frac{\partial q_v}{\partial t} + \frac{\partial J}{\partial z} = 0, \qquad J = q_v v, \qquad m \frac{dv}{dt} = eE_z.$$

Note that the vector representation for J and v is no longer needed since they have only z-components. The $\mathbf{v} \times \mathbf{B}$ term does not appear in the last equation since we wrote the z-components of the equation, and since $\mathbf{v}$ is in the z-direction, the cross product will be perpendicular to z.

These equations divide into two sets. The first four contain only the four variables E_x, E_y, H_x, H_y. They are, in fact, the equations for the uniform plane wave in free space, which we investigated in Chapter 9, and which we shall not consider further here. We also have another four equations relating E_z, q_v, J, and v:

$$\epsilon_0 \frac{\partial E_z}{\partial t} + J = 0, \qquad \frac{\partial q_v}{\partial t} + \frac{\partial J}{\partial z} = 0,$$

$$J = q_v v, \qquad m \frac{dv}{dt} = eE_z.$$

We shall show that a solution of these leads to waves on the electron beam.

13–3 Waves on electron beams

In our previous investigations of waves we supposed all our quantities to be sinusoidally time varying. However, a moment's reflection will show that in the present context this is not an appropriate assumption. For example, we have already specified that the electron beam is moving in the z-direction, and thus $\mathbf{v}$ must have a steady component. We will suppose the signal to be a sinusoidally time-varying velocity superimposed on the steady component so that

$$v = v_0 + v_1 \cos(\omega t + \theta), \qquad \text{or} \qquad v = v_0 + \mathrm{Re}\,(\hat{v} e^{j\omega t}),$$

where $\hat{v}$ is the complex velocity, and $\hat{v} = v_1 e^{j\theta}$. Similarly J and q_v will consist of steady components and signal components,

$$J = J_0 + \mathrm{Re}\,(\mathcal{J} e^{j\omega t}) \qquad \text{and} \qquad q_v = (q_v)_0 + \mathrm{Re}\,(\hat{q}_v e^{j\omega t}).$$

We suppose that all four of the variables will take on this same form of variation. For a beam consisting of a uniform charge density moving with a uniform velocity with sinusoidally time-varying components superimposed on this steady motion, v_0, $(q_v)_0$, and J_0 are constants. The amplitudes of the time-varying components are $\hat{v}$, $\hat{q}_v$, and $\mathcal{J}$, and these quantities are complex. The fourth variable E_z can have only an ac-component, since if there were a steady field in the z-direction, the electrons would be accelerating. A region in which electrons move without acceleration is often referred to as a *drift* region. The form of the variation introduced for v, q_v, and J is exactly the same as that used in Section 12–2. We saw there how this form of variation results in the equations splitting into two, a dc-equation and an ac-equation. We also saw that by using the small-signal approximation in conjunction with this form of variation we were able to linearize a nonlinear equation. Examination of the above equations shows that some of these are nonlinear and thus the same procedure is applicable here. Using the above form for the variables and the small-signal approximation gives the following ac-equations:

$$j\omega\epsilon_0\mathcal{E}_z + \mathcal{J} = 0, \qquad j\omega\hat{q}_v + \frac{d\mathcal{J}}{dz} = 0, \qquad \mathcal{J} = \hat{q}_v v_0 + (q_v)_0\hat{v}.$$

The fourth equation requires a little more development. Remembering that v is a function of z and t, we have

$$\frac{dv}{dt} = \frac{\partial v}{\partial t} + \frac{\partial v}{\partial z}\frac{\partial z}{\partial t} = \frac{\partial v}{\partial t} + v\frac{\partial v}{\partial z}\cdot$$

The same procedure as above then gives the equation

$$j\omega\hat{v} + v_0\frac{d\hat{v}}{dz} = \frac{e}{m}\mathcal{E}_z.$$

We now have four equations for the four unknowns $\hat{v}$, $\hat{q}_v$, $\mathcal{J}$, and $\mathcal{E}_z$. Eliminating $\hat{q}_v$ between the second and the third gives

$$\frac{d\mathcal{J}}{dz} + \frac{j\omega}{v_0}\mathcal{J} = \frac{j\omega(q_v)_0}{v_0}\hat{v}.$$

Similarly eliminating $\mathcal{E}_z$ between the first and the fourth gives

$$\frac{d\hat{v}}{dz} + \frac{j\omega}{v_0}\hat{v} = -\frac{e}{j\omega\epsilon_0 m v_0}\mathcal{J} = j\frac{e}{\omega\epsilon_0 m v_0}\mathcal{J}.$$

We thus have two simultaneous first-order differential equations in $\hat{v}$ and $\mathcal{J}$. The right-hand side of these equations becomes an exact differential if we multiply through by $e^{j\omega z/v_0}$:

$$\frac{d\mathcal{J}}{dz}e^{j\omega z/v_0} + \frac{j\omega}{v_0}\mathcal{J}e^{j\omega z/v_0} = \frac{d}{dz}(\mathcal{J}e^{j\omega z/v_0}) = j\frac{\omega(q_v)_0}{v_0}\hat{v}e^{j\omega z/v_0}.$$

If we let $\omega/v_0 = \beta_e$, $\mathcal{J}e^{j\omega z/v_0} = \mathcal{J}e^{j\beta_e z} = \mathcal{J}'$, and $\hat{v}e^{j\beta_e z} = \hat{v}'$, then

$$\frac{d\mathcal{J}'}{dz} = j\beta_e(q_v)_0\hat{v}'.$$

Similarly from the second equation we obtain

$$\frac{d\hat{v}'}{dz} = j\frac{e}{\omega\epsilon_0 m v_0}\mathcal{J}'.$$

Hence eliminating $\hat{v}'$ gives us

$$\frac{d^2\mathcal{J}'}{dz^2} = j\beta_e(q_v)_0 j\frac{e}{\omega\epsilon_0 m v_0}\mathcal{J}' = -\frac{e(q_v)_0}{\epsilon_0 m v_0^2}\mathcal{J}' = -\beta_p^2\mathcal{J}',$$

where B_p^2 is given by

$$\beta_p^2 = \frac{e(q_v)_0}{\epsilon_0 m v_0^2}.$$

The solutions for $\mathcal{J}'$ are therefore of the form $e^{\pm j\beta_p z}$, and for $\mathcal{J}$, $e^{-j(\beta_e \pm \beta_p)z}$. Thus the solutions for J are of the form $\cos\,[\omega t - (\beta_e \pm \beta_p)z]$, which represents a propagating wave. Therefore we see that there are two possible waves which can propagate on the beam, the phase constants being $\beta_e \pm \beta_p$. The phase constant $\beta_e = \omega/v_0$ is known as the electronic phase constant, and $\beta_p = \omega_p/v_0$ is the plasma phase constant, ω_p being the plasma frequency given by $\omega_p^2 = e(q_v)_0/\epsilon_0 m$. We obtain the velocity from the relation

$$\hat{v}' = \frac{1}{j\beta_e(q_v)_0}\frac{d\mathcal{J}'}{dz}.$$

Taking the solution $e^{+j\beta_p z}$ and normalizing the amplitude of $\mathcal{J}$ to unity, we have

$$\hat{v}' = \frac{\beta_p}{\beta_e(q_v)_0}\, e^{j\beta_p z},$$

and thus

$$\hat{v} = \frac{\beta_p}{\beta_e(q_v)_0}\, e^{-j(\beta_e-\beta_p)z}.$$

The charge density is given by

$$\hat{q}_v = -\frac{1}{j\omega}\frac{d\mathcal{J}}{dz} = \frac{\beta_e - \beta_p}{\omega}\, e^{-j(\beta_e-\beta_p)z}.$$

For the second wave we have

$$\mathcal{J} = e^{-j(\beta_e+\beta_p)z},$$

$$\hat{v} = -\frac{\beta_p}{\beta_e(q_v)_0}\, e^{-j(\beta_e+\beta_p)z},$$

$$\hat{q}_v = \frac{\beta_e + \beta_p}{\omega}\, e^{-j(\beta_e+\beta_p)z}.$$

We obtain the phase velocities of these waves from the ratio of frequency to phase constant, and so

$$v_p = \frac{\omega}{\beta_e \pm \beta_p} = \frac{v_0}{1 \pm \omega_p/\omega}.$$

Plotting this relation as a function of v_p against ω gives the graph shown in Fig. 13-1. The upper sign gives branch 1 of this curve; the wave has a phase velocity less than the beam velocity for all frequencies and is known as the

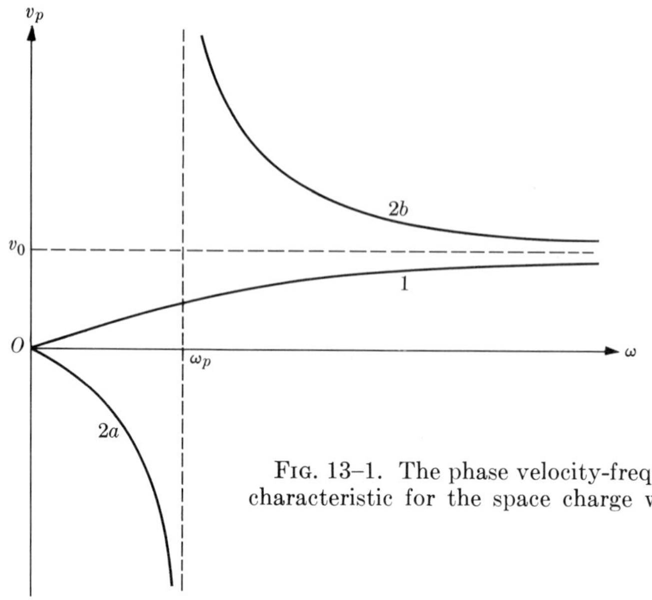

FIG. 13-1. The phase velocity-frequency characteristic for the space charge waves.

slow wave. The lower sign gives branch 2 of the curve. For frequencies above the plasma frequency, the wave has a velocity greater than the beam velocity and is known as the *fast wave.* Below the plasma frequency the phase velocity is in the opposite direction to the beam velocity.

We can readily obtain the group velocity from the above results. Since $1/v_g = d\beta/d\omega$ and $\beta = \beta_e \pm \beta_p = \omega/v_0 \pm \omega_p/v_0$, we have $1/v_g = 1/v_0$ or $v_g = v_0$. Thus the group velocity is equal to the beam velocity for both waves.

We have seen above that the charge density, by virtue of the form of its variation $\cos [\omega t - (\beta_e \pm \beta_p)z]$, propagates down the beam. At any given instant there is a sinusoidal variation of charge density with z, and the electrons are in bunches, which propagate down the beam in much the same way as an acoustic wave. The waves we have derived are known as *space charge waves.* The energy transported by the beam is carried in the form of kinetic energy by the electrons. Let us consider how this energy is modified by the presence of a fast space charge wave on the beam. We derived the formulas for the velocity and charge density above. The relation between the complex amplitudes is given by

$$\hat{v} = \frac{\omega}{(q_v)_0} \frac{\beta_p}{\beta_e(\beta_e - \beta_p)} \hat{q}_v,$$

and since ω, $(q_v)_0$, β_e, and β_p are real, the time-varying quantities v and q_v will be in phase provided that $\beta_e > \beta_p$, that is, if the signal frequency is greater than the plasma frequency. This means that at those points where the charge density is greater than it is in the absence of the wave, the velocity is also greater. Conversely where the charge density is lower, the velocity is lower.

Thus there are more electrons with greater than average energy and fewer with less than average energy, so a beam with a fast space charge wave on it carries greater energy than the unmodulated beam. For the slow wave, the charge density and the velocity are 180° out of phase so that fewer electrons will have greater than average energy and a larger number will have less than average energy. A beam with a slow space charge wave on it therefore carries less energy than the unmodulated beam.

The analysis above has been carried out for an infinite beam. Obviously in practice finite beams must be used. This case can also be analyzed, and it requires a step corresponding to the one we took in going from the waves in the infinite medium in Chapter 9 to the waves in a finite medium bounded by a guide which we considered in Chapter 10. The analysis is too complex to carry through here, but just as we found that an infinite number of higher-order modes can exist in the guide, so there are an infinite number of modes possible on the finite beam instead of just two.

We might also note in passing that the space charge mode is not the only type of mode possible on the infinite beam. Other types such as the cyclotron and synchronous modes can also exist. The electron motion in these cases is quite complex. The synchronous mode is perhaps the easier to consider. If we imagine a lawn sprinkler which throws out a jet of water and at the same time rotates, operating in the absence of gravity, then the jet of water is wound into a helix. Note that any one droplet of water moves in a straight line. A snapshot of the water at any one instant would show the water in a helix pattern. As time progresses this whole pattern propagates. The motion is analagous to the motion of the electrons in the synchronous mode. If we imagine the helical pattern of electrons to be fixed on a cylindrical frame, then the motion in the cyclotron mode would be obtained by rotating the frame at the cyclotron frequency, in addition to translating it parallel to its axis. The frequency of the wave excited on the beam determines how tightly wound the helix is (i.e., its pitch), and the translational velocity is equal to the beam velocity. A complete mathematical analysis of these modes has been made but is beyond the scope of the present work.

13-4 Waves on periodic structures

We saw in Chapter 10 that the hollow waveguide supports propagation modes whose phase velocities are greater than the velocity of light in the medium filling the waveguide. A periodic structure will support waves with phase velocities which are less than the velocity of light, and for this reason, it is often referred to as a *slow-wave structure*. By a periodic structure we mean one whose properties vary periodically with length, for example, a crystal. Here, however, we shall confine our discussion to structures which are periodic in one direction only, unlike the crystal which is periodic in all three dimensions. The periodicity may either be in the medium, such as a periodic variation of the permittivity with distance down the guide, or in the boundary, such as in the case of a corrugated waveguide. The latter are obviously more easily constructed and are the ones used in practice. A complete field analysis of these devices is quite complicated and will not be attempted here. However, some general remarks may be made which will enable us to see some of their properties.

Let us consider a structure which has a periodicity in the z-direction and suppose that the period is L. If we have a wave propagating down the guide in the z-direction, and if the variation of the complex field with z is given by $f(z)$, then it may be shown that

$$f(z + L) = e^{-\gamma L} f(z),$$

where γ is the propagation constant. This result follows from Floquet's theorem. Its proof is somewhat involved and will not be attempted here. Considering the lossless case, we can write this as

$$f(z + L) = e^{-j\beta_0 L} f(z).$$

We now define a new function $g(z) = e^{j\beta_0 z} f(z)$, and substituting, we have

$$g(z + L)e^{-j\beta_0(z+L)} = e^{-j\beta_0 L}e^{-j\beta_0 z}g(z),$$

$$\therefore g(z + L) = g(z),$$

and the function g is periodic in z with period L. We can therefore represent it as the Fourier series

$$g(z) = \sum_{-\infty}^{\infty} C_n e^{-j(2n\pi/L)z},$$

and the z-variation of the field $f(z)$ is given by

$$f(z) = \sum_{-\infty}^{\infty} C_n e^{-j\beta_0 z} e^{-j(2n\pi/L)z} = \sum_{-\infty}^{\infty} C_n e^{-j\beta_n z},$$

where $\beta_n = \beta_0 + 2\pi n/L$. Let us consider, for example, the z-component of the electric field. Then

$$\mathcal{E}_z = \sum_{-\infty}^{\infty} C_n e^{-j\beta_n z},$$

and E_z may be written

$$E_z = \sum_{-\infty}^{\infty} A_n \cos\left[(\omega t - \beta_n z) + \phi_n\right].$$

This represents a superposition of a series of propagating waves, all of which have the same frequency ω but different phase constants and therefore different phase velocities. We consider an elementary periodic structure consisting of a series of thin parallel slots in the x-direction cut in a block of conducting material, the whole structure being infinite in the x- and z-directions (Fig. 13–2). In general, E_z will be a function of x, y, z, and t, and C_n will thus be a function of x and y. In this case, since the structure is infinite in the x-direction, we may suppose C_n to be a function of y only. Considering the field at $y = 0$ at some instant of time (say $t = 0$), we can see that the variation of E_z with z will be quite complicated. It must be zero between the slots (such as the region ab), but may have a finite and variable magnitude across the slot (such as across bc). It is to synthesize this complicated variation with z that we need the Fourier series. We are thus doing a Fourier analysis in space, and the individual terms $\cos(\omega t - \beta_n z)$ are known as *space harmonics* (or *Hartree harmonics*). Note that they are not harmonics as we usually think of them, since they all have the same frequency. Here it is the wavelengths that are different. The group velocity is given by $d\omega/d\beta$, and thus we have

$$\frac{1}{(v_g)_n} = \frac{d\beta_n}{d\omega} = \frac{d\beta_0}{d\omega} = \frac{1}{(v_g)_0}.$$

Hence all the space harmonics have the same group velocity, and all carry energy at the same rate.

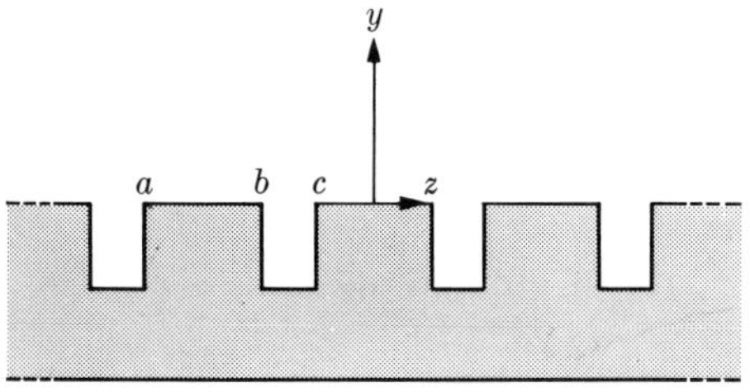

FIG. 13–2. A periodic structure.

In Section 10–3 we developed the idea of the $\omega\beta$-diagram and noted that this is one of the useful ways of investigating the more complicated structures. To obtain the $\omega\beta$-diagram, we need a complete analysis of the structure under investigation. However, we can determine some general properties of the diagram without this. Let us suppose for a given frequency we have obtained the phase constant β_0. This is one point on the curve. There will also be a phase velocity $\beta_0 + (2\pi/L)$ at this frequency, which constitutes another point on the curve. By extension of the argument to include all the harmonics, we see that the $\omega\beta$-diagram must be periodic in β, with period $2\pi/L$. From the symmetry of the structure we can also say that for each wave traveling in the $+z$-direction with a particular phase velocity, there must be a corresponding wave traveling in the $-z$-direction with the same phase velocity. The $\omega\beta$-diagram must therefore be an even function of β. Another point which a more detailed investigation shows is that a periodic structure exhibits alternating pass and stop bands. We can see in a general way how this occurs. The slots in the block can be considered

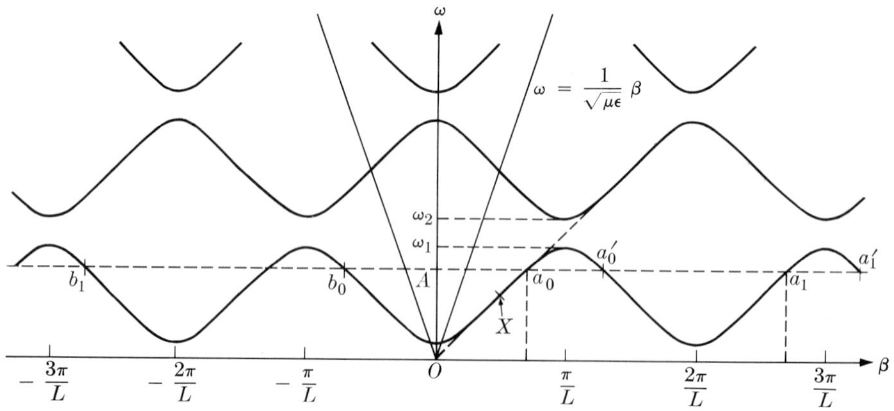

FIG. 13-3. The $\omega\beta$-diagram for a periodic structure.

as discontinuities, and from each one there will be a backscattered wave. When the frequency is such that all backscattered waves from the slots are in phase, they will reinforce each other, a strong reflected wave will form, and transmission cannot take place. This will occur when the distance between the slots is one half-wavelength. If the frequency is increased, the backscattered waves are no longer in phase, and no strong reflected wave is formed; hence transmission can again take place. On increasing the frequency still further one will reach a point where the spacing of the slots is one wavelength and the backscattered waves are again in phase. This frequency is the center of the second stop band, and we see that there will be stop bands centered on frequencies which correspond to slot spacings of an integral number of half-wavelengths with transmission bands between them. Thus a complete $\omega\beta$-diagram for some periodic structure would be as shown in Fig. 13-3. If we choose to work at some frequency, say A, then the points $a_0, a_1, \ldots$ give the series of phase constants β_n. We saw in Section 10-3 that the phase velocity is the slope of the line drawn to the point from the origin and that the group velocity is the slope of the curve. Thus the points $a_0, a_1, \ldots$ represent the space harmonics, each with a different phase velocity and the same group velocity. The line $\omega = \beta/\sqrt{\mu\epsilon}$ is plotted on the diagram, and points below this line represent waves with a phase velocity less than the velocity of light. Thus in this diagram all the space harmonics are slow waves. The points $b_0, b_1, \ldots$ represent the same waves traveling in the opposite direction. At the frequency corresponding to the point A we also have the series of intersections $a_0', a_1', \ldots$ The slope of the curve at these points is negative, and so the group velocity is negative. The phase velocity is still positive, however. These waves, which have their phase and group velocities in opposite directions, are known as *backward waves*. The region of frequencies from ω_1 to ω_2 constitutes the stop band, and no propagation is possible between these two limits. Note that at the cutoff frequencies ω_1 and ω_2, the group velocity goes to zero. If we consider the point X on the curve, we see that the broken line drawn from the origin coincides with the $\omega\beta$-diagram over a considerable range, which means that the phase velocity is approximately constant over a wide frequency range. The significance of this fact will be taken up in the next section.

There are various slow-wave structures in use today. One of the most common is the helix, which is suited to low-power operation. For higher powers a circular waveguide with washers spaced at equal intervals may be used. Many other configurations have also been employed.

13-5 Traveling wave tubes

The traveling wave tube is based on the interaction of the waves on the electron beam and the slow-wave structure. A theoretical analysis of the devices can be made using the method of coupled modes. Such an analysis is beyond the scope of the present work, but a general description of the operation of the traveling wave tube can be given. It is helpful to think in terms of an analogy. Let us consider two tuned circuits, one of which is being driven. If we are to obtain interaction between the two circuits, "interaction" meaning, for example, a marked change in the tuning characteristic or a transfer of energy from one circuit to the other, then two conditions must be fulfilled. First, the two circuits must be coupled, which may be accomplished by positioning the inductance of the second circuit in such a way that the magnetic field from the first inductance is linked with it and thus can induce oscillations in the second circuit. Second, the two circuits must resonate at approximately the same frequency. We know that there is little response in the secondary of a transformer unless it is correctly tuned. To obtain interaction between the waves on an electron beam and on a circuit, two similar conditions must be fulfilled.

We consider a traveling wave tube consisting of a helix with an electron beam traveling down its axis (Fig. 13-4). An input wave travels down the waveguide and excites the cavity. The helix is in the field produced by the cavity, and this in turn excites the waves on the helix. All the space harmonics corre-

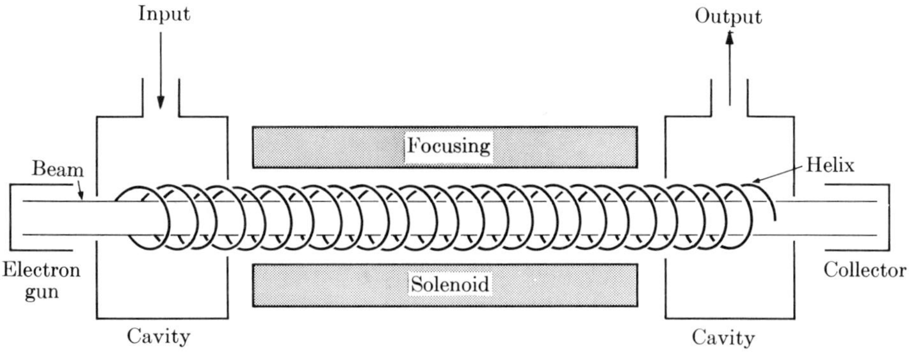

FIG. 13-4. A traveling wave tube.

sponding to the particular frequency will be excited so that the boundary conditions for the helix are satisfied. Furthermore, to satisfy the boundary condition, the resulting field must contain both TE- and TM-components. The TM-field has a longitudinal component of electric field, and this can excite the space charge modes on the beam, which as we have seen, also has a longitudinal electric field. Thus our first condition is fulfilled. There are suitable fields to cause coupling between the circuit mode and the space charge mode. The second condition in the case of propagating modes is that the phase velocities shall be approximately the same (corresponding to tuning in coupled circuits). We can see this in a general way. Let us suppose that a field due to the circuit begins to create a disturbance on the beam. Both of these will now propagate. If they propagate together, then the action of the circuit field on the beam stays the same and the disturbance on the beam will grow. If, however, the velocities are different, then the small disturbance on the beam will come under the influence of different parts of the field on the circuit. These will periodically reverse and cancel the original disturbance. This is a rather crude explanation, but serves to illustrate the point. To satisfy the second condition, we must adjust the design of the slow-wave structure and the beam voltage so that the phase velocities are equal. This explains the need for a slow-wave structure. Since the beam is composed of electrons which have mass, it cannot have a velocity greater than the velocity of light and, in general, will have a much lower velocity. The space charge waves at microwave frequencies have a phase velocity roughly the same as that of the beam. Hence the phase velocity on the circuit must also be less than the velocity of light, and a slow-wave structure is therefore needed. The phase velocity on the helix is not unique. Interaction will be obtained, however, between a particular space harmonic and the space charge wave if the phase velocities are nearly equal. If, due to the interaction, the magnitude of this space harmonic is increased, then for the boundary conditions to remain satisfied all the other space harmonics must increase in the same proportion. Thus the adjusting of the phase velocity of the beam wave to any one of the space-harmonic velocities results in interaction between the space charge wave on the beam and the wave on the circuit.

To explain the operation of the traveling wave tube as an amplifier we consider a forward mode on the circuit (phase and group velocities in the same direction and also in this case in the same direction as the beam) coupled to a slow space charge wave. The power is fed into the circuit at the electron-gun end of the tube. Owing to the interaction between the circuit and the beam, the slow wave is excited on the beam. We saw in Section 13–3 that when the beam has a slow wave on it, less energy is carried. The interaction thus reduces the energy on the beam, and since the law of conservation must be obeyed, this must be transferred to the circuit. The wave on the circuit therefore increases in amplitude. Both waves grow as they progress down the tube, the growing slow wave on the beam producing a continual reduction of energy in the beam, and the growing wave on the circuit producing a continual increase in the circuit energy. At the end of the tube the amplified wave on the circuit is taken out via a cavity and a waveguide, and the electron beam is collected. Thus we get increased signal energy at the output at the expense of the beam energy. This may be compared with increased signal energy at the plate of the vacuum-tube amplifier at the expense of B_+ energy.

We saw in the last section that the slow-wave structure can be designed so that the phase velocity of a particular harmonic is constant over a wide frequency range. We shall therefore fulfill the condition for coupling over a wide frequency range if the tube is designed so that coupling takes place with this harmonic. The traveling wave tube can therefore be designed to be a broad-band amplifier.

Finally let us consider the backward wave devices. In this case energy is put into the circuit at the collector end. The excited backward wave then carries energy back in the direction of the gun, but has a phase velocity in the same direction as the beam. Again we couple to the slow wave. We have no mechanism for producing the electron beam with the space charge wave already on it, and so we must suppose the magnitude of the slow wave to be zero at the gun and increase to some value at the collector end of the tube. There will therefore be less energy in the collected beam, and so again energy must be transferred to the circuit. As the waves progress together from the point of view of their phase velocities from the gun to the collector end of the tube, the magnitude of the slow wave increases and that of the circuit wave decreases. Thus the energy carried by the beam in the $+z$-direction decreases, and the energy carried by the circuit in the $-z$-direction also decreases, the two balancing each other so that energy conservation is satisfied. The device therefore works as a backward wave amplifier (BWA).

The detailed analysis shows that as z increases, the magnitude of the circuit wave decreases until at some particular value of z it becomes zero. If the tube is made this long, then the input is zero, i.e., there is no input signal. There is nevertheless an output signal, and hence the device is working as an oscillator. This is the backward wave oscillator (BWO), which is widely used in microwave engineering.

Problems

13–1 Using the relations derived in Section 13–3, calculate an expression for $\mathcal{E}_z$ and hence for E_z.

13–2 Derive the phase velocity and the group velocity for the E_z-wave obtained in Problem 13–1.

13–3 Suppose in a particular problem that the plasma frequency is higher than the wave frequency. How will this affect the arguments given in Section 13–3 concerning the energy carried by the space charge waves?

13–4 Show that the expression derived for $\mathbf{E}$ in Problem 13–1 satisfies Maxwell's equations.

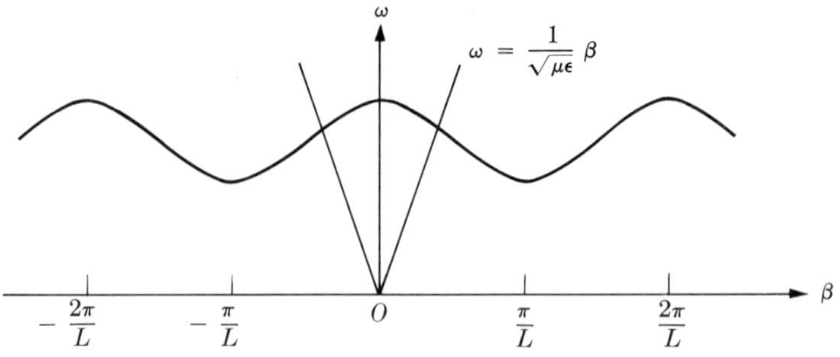

FIG. 13–5. An $\omega\beta$-diagram.

13–5 Verify that the expressions obtained in the text for $\hat{q}$ and $\mathcal{J}$ satisfy the continuity equation.

13–6 Explain why the space charge waves derived in Section 13–3 can be classed as uniform plane waves.

13–7 How would you describe the polarization of these waves?

13–8 If a particular electron in an electron beam with a fast space charge wave on it is at the point $z = 0$ and has velocity v_0 at time $t = 0$, derive an expression for the position of the electron as a function of time.

13–9 Given that the accelerating voltage for an electron beam is 1 kv, calculate the electron velocity. Assuming that the electron density in the beam is 10^{14} electrons/m^3, calculate the phase velocities of the fast and slow space charge waves [$\epsilon_0 = (1/36\pi) \times 10^{-9}$ f/m, $m = 9.1 \times 10^{-31}$ kg, $e = 1.6 \times 10^{-19}$ coul].

13–10 The $\omega\beta$-diagram for a particular periodic structure is shown in Fig. 13–5. By labeling your diagram show which region you would work in to obtain a broadband device. What is the group velocity in this region?

CHAPTER 14

Introduction to Plasma Physics

14–1 Introduction

We saw in earlier chapters how electric charges and currents in the presence of electric and magnetic fields give rise to mechanical forces. If these charges and currents are carried on a material body, then the mechanical forces will act on the mass of the body and produce accelerations. The resulting motion of the body will be governed by the laws of mechanics. We have already considered some of the effects which material media have on electric and magnetic fields. In Chapter 8 we saw how the fields are modified by the polarization and magnetization of the medium, and we saw in Chapter 9 how the permittivity and permeability of the medium affects waves propagating in it. In Chapter 11 we considered the motion of individual particles in fields, and in Chapter 12 we saw how this motion affects propagating waves. In the last chapter we considered the case of the electron beam where the mutual effect of the electrons on each other was taken into account (i.e., the effect of space charge). However, although we have been using the laws of mechanics in the treatment of these cases, we have not thus far allowed for any bulk motion of the material containing the charges. This is the final extension of the treatment given here and forms the subject matter of this chapter. We begin by making a further investigation of the motion of charged particles, a subject first taken up in Chapter 11.

14–2 The motion of charged particles

The fundamental law of mechanics governing the motion of a charged particle may be written

$$m\frac{d\mathbf{v}}{dt} = q(\mathbf{E} + \mathbf{v} \times \mathbf{B}).$$

Let us first consider the case where $\mathbf{E} = 0$. It is convenient to resolve the velocity into the two components, $v_{||}$ parallel to the field, and $v_\perp$ perpendicular to it. The $v_\perp$-component produces a force $v_\perp B$ acting at right angles to the velocity and thus produces a circular motion of radius

$$\frac{mv_\perp^2}{a} = qv_\perp B,$$

so that

$$a = \frac{mv_\perp}{qB},$$

and the angular frequency is given by $\omega = qB/m$. Superimposed on this circular motion is the velocity component parallel to the field, $v_{||}$, and the general motion of the particle is therefore a helix. We now introduce an **E**-field perpendicular to **B**. If we consider a positively charged particle, then when this is traveling in the direction of **E** it is accelerating, and the radius of curvature of its trajectory increases and reaches a maximum at the limit of travel in this direction. Conversely the curvature decreases as the particle travels against the field. Hence we see qualitatively that the particle must drift in a direction perpendicular to both the **E**- and **B**-fields (Fig. 14-1). This is known as the $\mathbf{E} \times \mathbf{B}$ drift, which can be calculated from the equation of motion by means of the substitution

$$\mathbf{v} = \mathbf{v}' + \frac{\mathbf{E} \times \mathbf{B}}{B^2},$$

Hence we have

$$m\frac{d\mathbf{v}'}{dt} = q\left[\mathbf{E} + \mathbf{v}' \times \mathbf{B} + \frac{(\mathbf{E} \times \mathbf{B}) \times \mathbf{B}}{B^2}\right].$$

In this case, since $\mathbf{E} \cdot \mathbf{B} = 0$,

$$(\mathbf{E} \times \mathbf{B}) \times \mathbf{B} = (\mathbf{B} \cdot \mathbf{E})\mathbf{B} - (\mathbf{B} \cdot \mathbf{B})\mathbf{E} = -B^2\mathbf{E},$$

$$m\frac{d\mathbf{v}'}{dt} = q\mathbf{v}' \times \mathbf{B}.$$

Thus $\mathbf{v}'$ is just the motion of the particle about the **B**-lines which occurs in the absence of the electric field. Superimposed on this is the velocity $(\mathbf{E} \times \mathbf{B})/B^2$, which is the drift perpendicular to **E** and to **B**. In the general case, where **E** is not perpendicular to **B**, we can resolve it into the components $E_\perp$ and $E_{||}$, as we did with velocity. The perpendicular component then produces a drift $v_D = E_\perp/B$, and the parallel component produces an acceleration along the **B**-lines. We may note in passing that the drift velocity is independent of the mass, charge, and velocity of the particles.

An electric field produces drift velocity because it exerts a force on the particle perpendicular to the magnetic field, and such a force will produce a drift no matter how it arises. A case of practical interest is the force produced by a gravitational field acting on the mass of the particle. If we suppose that the component of the gravitational field perpendicular to the magnetic field is $g_\perp$, then the force on the particle is $mg_\perp$, and the drift velocity will be $mg_\perp/qB$. Here the drift velocity depends on the charge and is in opposite directions for oppositely charged particles. The drifts resulting from gravitational fields therefore produce a current.

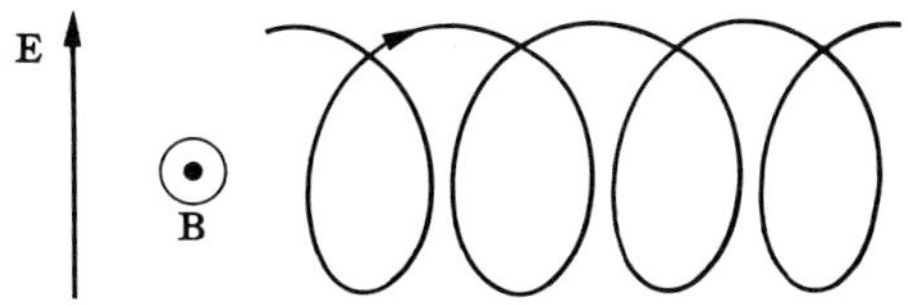

FIG. 14–1. The **E** × **B** drift.

So far we have supposed the magnetic field to be constant. An inhomogeneity in **B** can also produce a drift which is known as a *gradient drift.* Let us consider the situation in which **B** has only a z-component, which is a function only of x, that is, $\mathbf{B} = \mathbf{u}_z B_z(x)$. To see qualitatively how this drift arises, we assume that there is a discontinuity in the field at $x = 0$ and that the field is larger for $x > 0$. Hence for $x > 0$, the particle describes an orbit of smaller radius, and as a result, we have a drift in the y-direction (Fig. 14–2). An analysis of this motion shows that the drift velocity for this particular situation may be written

$$v_D = \frac{av_\perp}{2B_0}\frac{\partial B}{\partial x},$$

and for the general case,

$$\mathbf{v}_D = \frac{av_\perp \mathbf{B} \times \operatorname{grad} \mathbf{B}}{2B^2}.$$

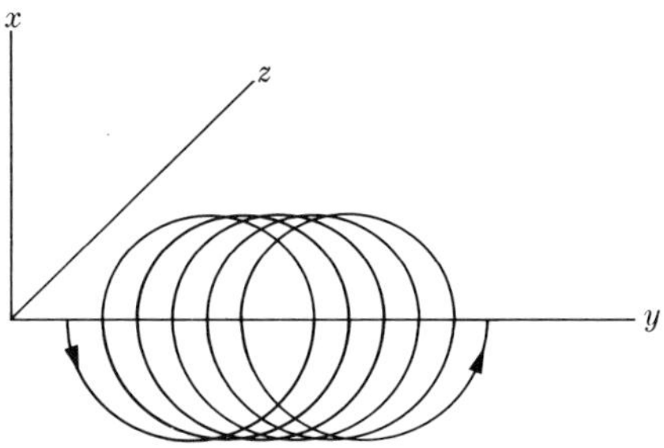

FIG. 14–2. The gradient drift.

Drift also arises when the lines are curved. A particle moving along a curved **B**-line experiences a centrifugal force which produces a drift in the same manner as electric and gravitational fields do. This is the *curvature drift.*

We shall now consider the behavior of a particle moving into a region where the **B**-lines are converging. This will mean that **B** is increasing in the direction of motion of the particle, and so the radius of the orbit decreases, as shown in Fig. 14–3. The converging of the **B**-lines also means that there is a component of **B** in the plane of the orbit. We suppose that the central **B**-line is the z-axis. For $z = 0$, there is then a component of **B** in the $-\rho$-direction (using conven-

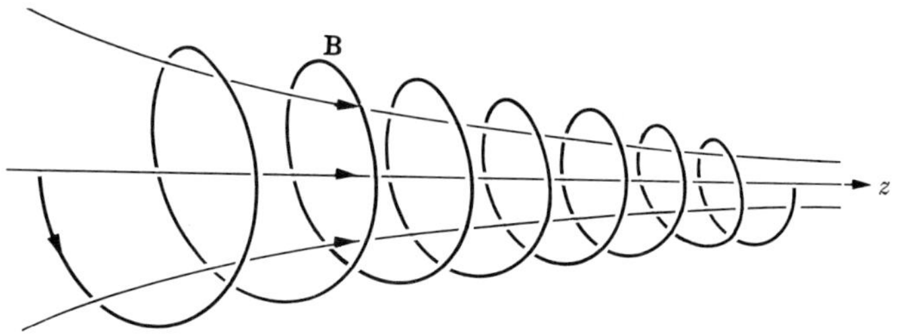

FIG. 14–3. Particle drift in converging **B**-lines.

tional notation for cylindrical polar coordinates), i.e., perpendicular to the z-axis and directed toward it. This will result in a force acting on the particle parallel to the z-axis, and it is readily verified that this acts in the $-z$-direction and thus retards the drift of the particle in the z-direction. We will treat this case analytically. We know from our fundamental equations that div $\mathbf{B} = 0$. If $\mathbf{B}$ has components in the ρ- and z-directions only, as in this case, then we can write (Appendix A-5)

$$\operatorname{div} \mathbf{B} = \frac{1}{\rho}\frac{\partial}{\partial \rho}(\rho B_\rho) + \frac{\partial B_z}{\partial z} = 0,$$

and hence

$$B_\rho = -\frac{1}{\rho}\int \rho \frac{\partial B_z}{\partial z}\, d\rho.$$

If we assume that $\partial B_z/\partial z$ is constant over the particle orbit, and that $B_\rho \ll B_z$ so that $B_z \simeq B$, then we have

$$B_\rho = -\frac{\rho}{2}\frac{\partial B}{\partial z},$$

which gives B_ρ in terms of the change of B in the z-direction. If the radius of the particle orbit is a, then the radial field experienced by the particle is

$$B_\rho = -\tfrac{1}{2}a\frac{\partial B}{\partial z}\cdot$$

The force in the z-direction acting on the particle is therefore given by

$$F_z = qv_\perp B_\rho = -qv_\perp \tfrac{1}{2}a\frac{\partial B}{\partial z},$$

where $v_\perp$ implies the velocity component perpendicular to z. Substituting for a gives

$$F_z = -qv_\perp \frac{1}{2}\frac{mv_\perp}{qB}\frac{\partial B}{\partial z}\cdot$$

A particle moving in a circle constitutes a current loop, and as we saw in Chapter 6, this is equivalent to a magnetic dipole. The current is given by the charge crossing a given point on the circle per unit time. The charge q makes $\omega/2\pi$ rev/sec, and the current is therefore $q\omega/2\pi$. The dipole moment is given by the current multiplied by the area and hence

$$\mu = \frac{q\omega}{2\pi}\pi a^2 = \frac{mv_\perp^2}{2B} = \frac{w_\perp}{B},$$

where $w_\perp$ is the kinetic energy due to $v_\perp$. We can therefore express F_z in terms of the dipole moment

$$F_z = -\mu\frac{\partial B}{\partial z}\cdot$$

From the equation of motion we have $F_z = m\, dv_z/dt$, and thus

$$mv_z \frac{dv_z}{dt} = \frac{d}{dt}\left(\tfrac{1}{2}mv_z^2\right) = -\mu v_z \frac{\partial B}{\partial z} = -\mu \frac{dB}{dt}.$$

Note that d/dt means the time rate of change of B as seen by the particle; i.e., the derivative is taken along the trajectory moving with the particle, not at a fixed point in space. Hence

$$\frac{d}{dt}\left(\tfrac{1}{2}mv_z^2\right) = -\mu \frac{dB}{dt}.$$

The total energy of the particle is $\frac{1}{2}mv_z^2 + \frac{1}{2}mv_\perp^2$, and from the law of conservation we have

$$\frac{d}{dt}\left(\tfrac{1}{2}mv_z^2 + \tfrac{1}{2}mv_\perp^2\right) = 0,$$

and since $\mu = mv_\perp^2/2B$, it can be seen that

$$\frac{d}{dt}\left(\tfrac{1}{2}mv_z^2\right) = -\frac{d}{dt}(\mu B).$$

Comparing the two expressions for $(d/dt)(\frac{1}{2}mv_z^2)$, we have

$$\frac{d}{dt}(\mu B) = \mu \frac{dB}{dt},$$

and thus we obtain the result

$$\mu = \text{const.}$$

Hence when a particle is moving in a magnetic field which varies with position, it moves in such a way that the magnetic moment remains constant. It may also be shown that when a particle is moving in a magnetic field which varies with time, it too will move in such a way that the magnetic moment remains constant.

Returning to the motion of the particle in a region of converging **B**-lines, we find that if θ is the angle between the z-axis and the velocity of the particle, then $v_\perp = v \sin\theta$ and $\frac{1}{2}mv_\perp^2 = \frac{1}{2}mv^2 \sin^2\theta$. Since $\mu = \text{const}$, it follows that $mv_\perp^2/2B = \text{const}$, and by the conservation law $\frac{1}{2}mv^2 = \text{const}$. Hence

$$\frac{B}{\sin^2\theta} = \text{const} = \frac{B_0}{\sin^2\theta_0},$$

where B_0 and θ_0 are the initial values. If B increases to the value $B_0/\sin^2\theta_0$, then $\theta = 90°$, and hence the velocity becomes purely transverse, and $v_z = 0$. If $\partial B/\partial z$ is still finite at this point, so that there is a B_ρ-component, then the interaction of this component with $v_\perp$ produces a force in the $-z$-direction, and the particle now begins to drift in this direction. The particle is effectively reflected from this point, and the device is known as a *magnetic mirror*. This

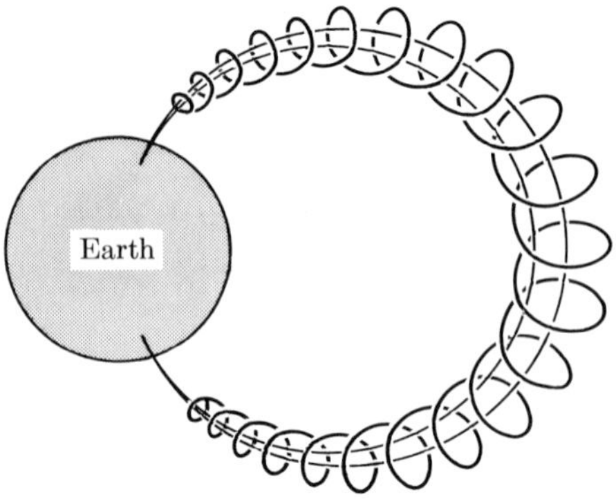

FIG. 14–4. A particle trapped by the field of the earth.

constitutes one principle for confining plasmas. The particles are confined radially by an axial magnetic field, and the ends of the container are "sealed off" with magnetic mirrors. The magnetic-mirror machines are of interest in connection with thermonuclear reactors. The radiation belts discovered by Van Allen are examples of particles confined in this manner on a cosmological scale. The magnetic field of the earth is roughly a dipole field, and thus the lines will be converging as they descend toward the surface of the earth. The particles are trapped by the magnetic field and are reflected back and forth between the two mirror points in the Northern and Southern Hemispheres (Fig. 14–4). The motion of these particles is quite complicated. We note that there is a gravitational field, a grad **B**, and curvature of the **B**-lines, so that in addition to the motion already described the particles will drift around the earth in the east-west direction.

14–3 Magnetohydrodynamics

Magnetohydrodynamics is concerned with the interaction between fields and matter when the behavior of the matter can be described by the macroscopic equations of hydrodynamics, and the motion of individual particles need not be considered. The law of conservation of mass can be expressed mathematically for a fluid as

$$\frac{\partial \rho}{\partial t} + \text{div } (\rho \mathbf{v}) = 0,$$

where ρ is the mass density and $\mathbf{v}$ the velocity of the fluid. The first term represents the rate of increase of mass within a small element, and the second term represents the flow of matter out of the element. We may compare this law with the equation of continuity which is an expression of the law of conservation of charge. The derivation of the two laws is similar.

We shall use the simplest form of the force equation,

$$\rho \frac{dv}{\partial t} = -\text{grad } p + \mathbf{J} \times \mathbf{B}.$$

The dimension of each term in this equation is force per unit volume. The left-hand side is mass per unit volume times acceleration. If the pressure is constant at all points in the fluid, then if we consider a volume element in the fluid, there will be no net force on it. If, however, the pressure varies from point to point (i.e., there is a pressure gradient), then there will be a force tending to move the element away from the region of higher pressure. This is the $-\text{grad}\ p$ term on the right-hand side. The final term is due to the interaction between currents in the fluid and the magnetic field.

We shall now investigate the phenomenon of magnetic pressure. From Maxwell's equations we have

$$\text{curl}\,\mathbf{H} = \mathbf{J} + \frac{\partial \mathbf{D}}{\partial t}\cdot$$

In good conductors, $\mathbf{J}$ is large compared to $\partial\mathbf{D}/\partial t$, and we may neglect the displacement current. We obtain

$$\mathbf{J} \times \mathbf{B} = \mu\,(\text{curl}\,\mathbf{H}) \times \mathbf{H}.$$

From the identity in the Appendix (Section A-3) we have

$$\mathbf{H} \times \text{curl}\,\mathbf{H} = \tfrac{1}{2}\,\text{grad}\,H^2 - (\mathbf{H}\cdot\text{grad})\mathbf{H}.$$

Hence

$$\rho\frac{\partial \mathbf{v}}{\partial t} = -\text{grad}\,p - \tfrac{1}{2}\mu\,\text{grad}\,H^2 + \mu(\mathbf{H}\cdot\text{grad})\mathbf{H}$$

$$= \mu(\mathbf{H}\cdot\text{grad})\mathbf{H} - \text{grad}\,(p + \tfrac{1}{2}\mu H^2).$$

Thus a term $\frac{1}{2}\mu H^2$ is introduced, which is due to the magnetic field and is equivalent to a hydrostatic pressure. This is known as the magnetic pressure.

Let us consider the simple case in which $\mathbf{H}$ has only a z-component ($H_x = H_y = 0$). Since $\text{div}\,\mathbf{H} = 0$, we must also have $\partial/\partial z = 0$. Hence $(\mathbf{H}\cdot\text{grad}) = 0$. In the steady state $\partial/\partial t = 0$, and so

$$\text{grad}\,(\tfrac{1}{2}\mu H^2 + p) = 0, \qquad \tfrac{1}{2}\mu H^2 + p = \text{const.}$$

Thus the sum of the magnetic and mechanical pressures is a constant. The mechanical pressure can therefore drop to zero, provided there is an accompanying increase in the magnetic pressure. This is the principle of plasma confinement by a magnetic field which is used in thermonuclear machines. We have seen that for a current of constant density flowing in a cylinder, the magnetic field increases linearly with radial distance from the axis to the edge of the current cylinder (Problem 6–8). If the current is carried by an ionized gas, then by the relation derived above the pressure of the gas must fall off with distance from the axis (on the assumption of constant current density). With appropriate design it is possible to produce the conditions shown in Fig. 14–5.

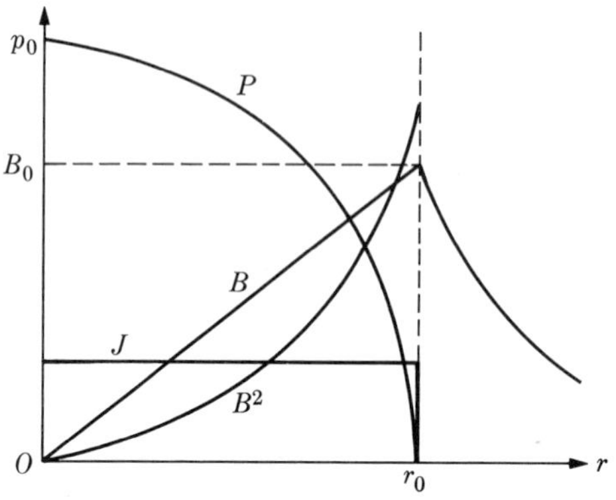

FIG. 14–5. Plasma confined by a magnetic field.

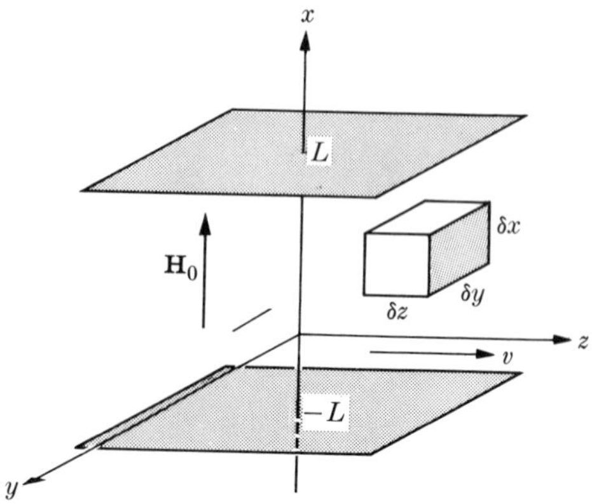

FIG. 14–6. Flow between parallel plates.

The plasma is now confined by its own magnetic field. This model is, of course, vastly oversimplified. We have assumed static conditions without any regard to the manner in which this situation is established. The assumption of uniform current is also not generally satisfied in practice. Furthermore, analysis shows that the condition above is not stable. Several types of instability occur which have been the subject of detailed investigation. The problem of confining a plasma has not yet been completely solved, and much work remains to be done in this area.

The currents induced in wires moving in magnetic fields constitute the principle of the dynamo, the standard means of generating the majority of our electric power today. There is, however, considerable interest in the magnetohydrodynamic generator, which uses a fluid instead of wires. As an example of the type of calculations involved we shall consider the simple case of a fluid flowing in the z-direction between two infinite plates parallel to the yz-plane (Fig. 14–6). The flow is produced by mechanical means, i.e., a pressure difference, and is opposed by viscous forces in the fluid. These arise because the velocity of the fluid at the plates which are stationary must be zero, and there is therefore a velocity gradient in the fluid. We also suppose that there is an externally applied magnetic field $\mathbf{H}_0$ acting in the x-direction. The motion of the fluid across the field produces an electric current in the y-direction. We shall first calculate the velocity $\mathbf{v}$ in terms of the pressure gradient, the applied field, and the constants of the fluid. We begin by developing the appropriate force equation. To simplify the problem we assume a steady state; i.e., the motion of fluid is not changing with time. There are therefore no accelerations in the system, and the net force acting must be zero. There are three force terms involved: the pressure gradient, the viscous force, and the electromagnetic force. For the pressure gradient we make the further simplifying assumption that it is constant throughout the fluid. We suppose it to act in the z-direction, and we write $-\text{grad}\ p = P$. To obtain the viscous force, let us consider the volume element shown. We know from the experimental law of viscous forces that the force per unit area is given by the coefficient of viscosity times the velocity gradient. For example, the force acting on the lower face of the element will be $\eta(dv/dx)\,\delta y\,\delta z$ and will act in the z-direction. The force on the upper face will be $\eta[dv/dx + (d/dx)(dv/dx)\,\delta x]\ \delta y\ \delta z$. Because of the velocity gradient, the fluid on one side of the element is moving more slowly than the element and will therefore tend to slow it down, and that on the opposite side is moving faster, tending to speed it up. The net force is thus the difference between the two terms above, and so the force per unit volume is $\eta(d^2v/dx^2)$. Since the plates are infinite in the y-direction, we will assume that the velocity does not depend on y, and if we specify that the fluid is incompressible, then the velocity is also independent of z. There is therefore only one component of the viscous force. The net force per unit volume due to the electric current is $\mathbf{J} \times \mathbf{B}$, and since $\mathbf{J}$ is in the y-direction and $\mathbf{H}_0$ in the x-direction, this force is in the $-z$-direction. Hence the force equation is

$$P + \eta \frac{d^2v}{dx^2} - \mu J H_0 = 0.$$

We now need an expression for J. This may be obtained for the general case as follows. The force on a moving charge is given by (Section 11–1)

$$\mathbf{F} = q_v(\mathbf{E} + \mathbf{v} \times \mathbf{B}).$$

The equivalent electric field experienced by the charge is therefore $\mathbf{E} + \mathbf{v} \times \mathbf{B}$, and if σ is the conductivity, we have $\mathbf{J} = \sigma(\mathbf{E} + \mathbf{v} \times \mathbf{B})$. There is no applied electric field, and thus $J = \sigma\mu v H_0$. Substituting for J in the force equation yields

$$\frac{d^2v}{dx^2} - \frac{\sigma\mu^2 H_0^2}{\eta} v + \frac{P}{\eta} = 0,$$

which gives an equation for v as a function of x. This may be solved by first making the substitution

$$v = v' + \frac{P}{\sigma\mu^2 H_0^2}\cdot$$

After matching the boundary conditions $v = 0$ at $x = \pm L$, we obtain

$$v = \frac{P}{\sigma\mu^2 H_0^2}\left[1 - \frac{\cosh\,(Mx/L)}{\cosh M}\right],$$

where $M = \mu H_0 L\sqrt{\sigma/\eta}$. This constant is frequently used in magnetohydrodynamic flow problems and is known as the *Hartmann number*. When M is large, the electric and magnetic quantities (μ, H_0, and σ) are large compared to the mechanical parameter η. Thus for large M we would expect magnetic effects to predominate, and for small M we would expect the results to approach those of hydrodynamics. Let us first consider M large. For $x > 0$ we may write

$$\frac{\cosh (Mx/L)}{\cosh M} \simeq \frac{e^{Mx/L}}{e^{M}} = e^{-M(1-x/L)},$$

and

$$v = \frac{P}{\sigma\mu^2 H_0^2}[1 - e^{-M(1-x/L)}].$$

For large M the exponential is negligible except near $x = L$. The velocity profile is then as shown in Fig. 14-7. For small M, $\cosh M \simeq 1 + \frac{1}{2}M^2$, and hence

$$v = \frac{P}{\sigma\mu^2 H_0^2}\left[1 - \left(1 + \frac{1}{2}\frac{M^2x^2}{L^2}\right)(1 - \tfrac{1}{2}M^2)\right] = \frac{PM^2}{2\sigma\mu^2 H_0^2}\left(1 - \frac{x^2}{L^2}\right)$$

or

$$v = \frac{PL}{2\eta}\left(1 - \frac{x^2}{L^2}\right),$$

and the parabolic profile is obtained as in classical hydrodynamics. The current produced is derived from the integral $\int\int J\,dx\,dz$; the integral with respect to x is taken from $-L$ to L, and if we consider a length of the plates w in the z-direction, the limits for z are 0 and w. Hence

$$\begin{aligned} I &= \int_{-L}^{L}\int_{0}^{w} J\,dx\,dz = \int_{-L}^{L}\int_{0}^{w} \sigma\mu H_0 v\,dx\,dz \\ &= \sigma\mu H_0 w \int_{-L}^{L} \frac{P}{\sigma\mu^2 H_0^2}\left[1 - \frac{\cosh (Mx/L)}{\cosh M}\right] dx \\ &= \frac{2PwL}{\mu H_0}\left(1 - \frac{\tanh M}{M}\right). \end{aligned}$$

The configuration considered here is, of course, not practical. It would be necessary in practice to have plates perpendicular to the y-direction to collect the current and connect it to the external circuit. This would confine the flow in the y-direction, which would have to be limited in any event, since in practice we cannot have infinitely large machines. This makes the problem much more complicated mathematically, and we shall not pursue the matter further.

As mentioned earlier, magnetohydrodynamic generators are of interest for power generation devices. Their advantage lies in the fact that the rotating parts of the conventional generators are replaced by a gas stream, and much higher temperatures can be used with consequent increase in thermal efficiency. The thermal energy of the fuel is converted into electrical energy without the

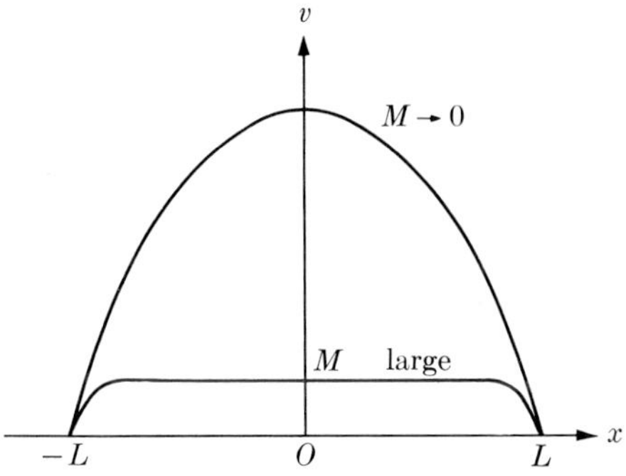

FIG. 14–7. The velocity profile for flow between parallel plates.

intermediary stage of conversion into mechanical energy, as is done in the steam turbines and rotating generators of the conventional power stations.

Just as there is a magnetohydrodynamic counterpart of the generator, so there is one of the motor. Let us suppose that we have a flow of ionized gas in the z-direction. If we apply a static magnetic field H_0 in the x-direction, then, as we have seen, there is an induced current in the y-direction. This current interacts with the steady field to produce a force density $\mathbf{J} \times \mathbf{B}$, which therefore acts in the $-z$-direction, and this opposes the flow of the fluid. However, if we apply an electric field in the $-y$-direction, we produce a current given by σE_0 in the $-y$-direction. If this exceeds the induced current in the $+y$-direction, then we produce a net current in the $-y$-direction, and the force $\mathbf{J} \times \mathbf{B}$ now acts in the $+z$-direction and accelerates the gas stream. We thereby produce a thrust by electromagnetic means. The induced current produced by the gas flow is $\sigma(\mathbf{v} \times \mathbf{B})$, and hence the condition for thrust is $\mathbf{E} > \mathbf{v} \times \mathbf{B}$. This process has obvious applications to propulsion devices. A typical motor of this type would consist of a chamber containing an electric arc into which the propellant is injected. This will raise the temperature of the gas to the point where it becomes conducting. The hot conducting gas then expands through a conventional rocket nozzle to produce a gas stream, which then passes through the crossed electric and magnetic fields, where it is accelerated. However, this is an application which is still in its infancy. Other applications of the foregoing ideas are the electromagnetic pumps used to circulate such fluids as molten sodium in nuclear reactors and the use of magnetic fields to control the boundary layer in high-speed flight, for example, during reentry into the atmosphere of the earth.

14–4 Radiation pressure

Let us consider a uniform plane wave traveling in the z-direction and incident on an infinite conducting plate perpendicular to the z-direction. The electric vector in the wave produces a force on the electrons in the conductor and thus causes a current which interacts with the magnetic field of the wave, thus producing a force. If the charge density in the conductor is q_v, then the force per unit volume acting on the conductor is $\mathbf{F}_v = q_v(\mathbf{E} + \mathbf{v} \times \mathbf{B})$. From Maxwell's equations we have $q_v = \operatorname{div} \mathbf{D}$ and $\mathbf{J} = \operatorname{curl} \mathbf{H} - \partial\mathbf{D}/\partial t$. Using the relation $\mathbf{J} = q_v\mathbf{v}$ and substituting, we have

$$\mathbf{F}_v = \mathbf{E} \operatorname{div} \mathbf{D} + (\operatorname{curl} \mathbf{H}) \times \mathbf{B} - \frac{\partial \mathbf{D}}{\partial t} \times \mathbf{B}.$$

Since

$$\frac{\partial}{\partial t}(\mathbf{D} \times \mathbf{B}) = \frac{\partial \mathbf{D}}{\partial t} \times \mathbf{B} + \mathbf{D} \times \frac{\partial \mathbf{B}}{\partial t} = \frac{\partial \mathbf{D}}{\partial t} \times \mathbf{B} - \mathbf{D} \times \operatorname{curl} \mathbf{E},$$

then

$$\mathbf{F}_v = \epsilon_0\mathbf{E} \operatorname{div} \mathbf{E} + \mu_0(\operatorname{curl} \mathbf{H}) \times \mathbf{H} - \epsilon_0\mu_0 \frac{\partial}{\partial t}(\mathbf{E} \times \mathbf{H}) - \epsilon_0\mathbf{E} \times \operatorname{curl} \mathbf{E}.$$

If we suppose the incident wave to be linearly polarized with $\mathbf{E}$ in the x-direction, then we have

$$\mathbf{F}_v = -\tfrac{1}{2}\mu_0 \frac{\partial}{\partial z}(H_y^2)\mathbf{u}_z - \tfrac{1}{2}\epsilon_0 \frac{\partial}{\partial z}(E_x^2)\mathbf{u}_z - \epsilon_0\mu_0 \frac{\partial}{\partial t}(\mathbf{E} \times \mathbf{H}).$$

The time-average energy flow is constant, and hence $(\partial/\partial t)(\mathbf{E} \times \mathbf{H}) = 0$, and so finally we obtain

$$(F_v)_z = -\frac{\partial}{\partial z}\left(\tfrac{1}{2}\epsilon_0 E_x^2 + \tfrac{1}{2}\mu_0 H_y^2\right).$$

The force is equal to the rate of decrease of stored energy with z. In a conductor energy is absorbed and thus decreases with z, and there is a force acting in the positive z-direction. We can calculate the magnitude of this force as follows. In the uniform plane wave $E_x/H_y = \sqrt{\mu_0/\epsilon_0}$, and hence the energy density associated with the wave may be written $w = \epsilon_0 E_x^2$. The Poynting vector $\mathbf{E} \times \mathbf{H} = E_x H_y \mathbf{u}_z$ in this case, and since $c = \sqrt{1/\mu_0\epsilon_0}$, we have $\mathbf{S} = c\epsilon_0 E_x^2 \mathbf{u}_z$. We therefore have the relation between the energy flow and the energy density $\mathbf{S} = w\mathbf{c}$. Hence the force per unit volume is given by

$$(F_v)_z = -\frac{1}{c}\frac{\partial S}{\partial z}.$$

If the wave is absorbed in the conductor in a short distance Δz, then $\mathbf{S}$ drops from its value in the incident wave $\mathbf{S}$ to zero in this distance, and the force per unit area in the z-direction is given by $\mathbf{S}/c$. This is the phenomenon of radiation pressure. Its existence has been demonstrated in the laboratory, and it has recently become of practical significance. A large light satellite such as the type used as a passive reflector in communications systems (e.g., the Echo satellite) experiences a considerable perturbation of its orbit as a result of radiation pressure from the rays of the sun. We can consider this phenomenon in terms of momentum. If we suppose the radiation to have momentum associated with it, then the absorption of the wave results in a change of momentum, and the rate of change produces the force. If the momentum per unit volume of the radiation is $\mathbf{g}$, then the rate of flow of momentum toward the absorbing plane is $\mathbf{g}c$ per unit area. If this radiation is completely absorbed, then the rate of change of momentum is also $\mathbf{g}c$, and this gives the force per unit area,

$$\mathbf{F}_s = \frac{\mathbf{S}}{c} = \mathbf{g}c, \qquad \text{or} \qquad \mathbf{g} = \frac{\mathbf{S}}{c^2}.$$

Let us consider this idea further. The energy W crossing a cross-sectional area ΔA perpendicular to the energy flow in time Δt is given by $S\,\Delta A\,\Delta t$. The momentum density associated with the radiation is S/c^2 or $W/(c^2\,\Delta A\,\Delta t)$. This radiation now occupies a volume $c\,\Delta A\,\Delta t$. If we assume that this radiation can have an equivalent mass m such that its momentum is mc, then the momentum density is $m/\Delta A\,\Delta t$. Hence equating the two, we obtain

$$\frac{m}{\Delta A\,\Delta t} = \frac{W}{c^2\,\Delta A\,\Delta t} \qquad \text{or} \qquad W = mc^2,$$

which is the well-known result in the theory of relativity, expressing the equivalence of mass and energy.

Problems

14–1 In Fig. 14–1, if the direction of the magnetic field is upward out of the page, what is the sign of the particle for the trajectory shown in the figure? Sketch the trajectory for the particle of opposite sign.

14–2 The drift velocity resulting from $\mathbf{E} \times \mathbf{B}$ is independent of the mass, charge, and velocity of the particle. Investigate whether the same is true for the detailed motion of the particle (i.e., the actual trajectory).

14–3 Investigate whether a gradient drift produces an electric current.

14–4 If the particle has kinetic energy $w_\perp$ due to a velocity component transverse to the magnetic field (the quantity used and defined in Section 14–2) and a kinetic energy $w_{||}$ defined in a similar fashion for the parallel component of velocity, what is the total kinetic energy? (This problem needs a careful interpretation of $w_\perp$ and $w_{||}$.) Is the equation relating kinetic energy and velocity linear? What is the significance of this in relation to the question posed above?

14–5 Check the qualitative argument given in Section 14–2 to show that converging B-lines can act as a magnetic mirror for particles with positive charge. Repeat for particles with negative charge.

14–6 In a given $\mathbf{B}$-field configuration forming a magnetic mirror, investigate whether the point of reflection depends on the magnitude of the charge on the particle, its mass, and its velocity.

14–7 By considering a suitable volume element, show that the force density acting on a current density $\mathbf{J}$ in a field $\mathbf{B}$ is given by $\mathbf{J} \times \mathbf{B}$.

14–8 Show that in a fluid in which the pressure varies with position, there is a force density $-\text{grad } p$. [*Hint:* Consider the forces on an infinitesimal volume element of cylindrical shape with its axis in the direction of grad p.]

14–9 In Section 14–3 we presented a simplified explanation of the mechanism by which a plasma can be confined by its own magnetic field. In the text a constant current density was assumed. Another model which has been used assumes that the current is confined to a thin layer at the surface of the cylinder. Investigate whether this can lead to confinement of the plasma.

14–10 In magnetohydrodynamic flow between two parallel plates, the velocity is approximately constant across the cross section of the flow for a large Hartmann number. Use this fact to get an approximate expression for the current. Show that the formula for I in the text gives the same result when M is made very large.

14–11 A uniform plane electromagnetic wave is incident normally on a poorly conducting region. Calculate the force density. By integrating this over a suitable volume, verify the result which was obtained in the text by assuming complete absorption.

APPENDIX

Vector Analysis

A–1 Introduction

Physical quantities such as mass or electric charge which have only a magnitude are known as *scalars*. The more complicated quantities which have direction as well as magnitude associated with them are *vectors*. In this text we use the bold-face characters (**A**) to denote a vector, and the italic type (A) to denote its magnitude. Thus A is a scalar. The magnitude symbol is also used on some occasions; for example, |grad V| means the magnitude of the gradient. Scalars are manipulated by the familiar laws of arithmetic. However, other rules must be defined for vectors.

A–2 Vector algebra

The operation of addition of vectors is carried out by the well-known parallelogram law. We can also resolve vectors into their components by the cosine law. For example, if the direction OL makes an angle θ with the vector **A**, then the magnitude of the component of **A** in the OL-direction is $A \cos \theta$. It is convenient to define directions by unit vectors. These are represented by **u**. We use a suffix to identify the particular direction, that is, $\mathbf{u}_l$ is the unit vector in the OL-direction, and $\mathbf{u}_x$, $\mathbf{u}_y$, $\mathbf{u}_z$ are the unit vectors in the rectangular coordinate system. The vector **A** may then be written in terms of its rectangular components

$$\mathbf{A} = A_x\mathbf{u}_x + A_y\mathbf{u}_y + A_z\mathbf{u}_z.$$

Note that A_x, A_y, and A_z are scalars. The law of addition in rectangular coordinates may be written

$$\mathbf{A} + \mathbf{B} = (A_x + B_x)\mathbf{u}_x + (A_y + B_y)\mathbf{u}_y + (A_z + B_z)\mathbf{u}_z.$$

The multiplication of a vector by a scalar can be expressed as

$$w\mathbf{A} = wA_x\mathbf{u}_x + wA_y\mathbf{u}_y + wA_z\mathbf{u}_z.$$

The scalar product between two vectors **A** and **B** is defined by the relation

$$\mathbf{A} \cdot \mathbf{B} = AB \cos \theta,$$

where θ is the angle between them. We stated above that the component of a vector in a given direction is obtained by the cosine law, and hence the magnitude of **A** in the $\mathbf{u}_l$-direction is $A_l = \mathbf{A} \cdot \mathbf{u}_l$. The scalar product may be written in rectangular coordinates,

$$\mathbf{A} \cdot \mathbf{B} = A_x B_x + A_y B_y + A_z B_z.$$

The vector product is defined to have magnitude $|\mathbf{A} \times \mathbf{B}| = AB \sin\theta$ and direction perpendicular to both **A** and **B**. The sense of the vector is such that a right-handed rotation is required to move **A** to the position of **B** through the smaller angle. It may be written in rectangular coordinates:

$$\mathbf{A} \times \mathbf{B} = \begin{vmatrix} \mathbf{u}_x & \mathbf{u}_y & \mathbf{u}_z \\ A_x & A_y & A_z \\ B_x & B_y & B_z \end{vmatrix}.$$

The various triple products are also of importance. We may write

$$\mathbf{A} \cdot \mathbf{B} \times \mathbf{C} = (A_x \mathbf{u}_x + A_y \mathbf{u}_y + A_z \mathbf{u}_z) \cdot \begin{vmatrix} \mathbf{u}_x & \mathbf{u}_y & \mathbf{u}_z \\ B_x & B_y & B_z \\ C_x & C_y & C_z \end{vmatrix},$$

$$\mathbf{A} \cdot \mathbf{B} \times \mathbf{C} = \begin{vmatrix} A_x & A_y & A_z \\ B_x & B_y & B_z \\ C_x & C_y & C_z \end{vmatrix}.$$

Since a determinant remains unchanged for a cyclic variation of its rows, we have

$$\mathbf{A} \cdot \mathbf{B} \times \mathbf{C} = \mathbf{C} \cdot \mathbf{A} \times \mathbf{B} = \mathbf{B} \cdot \mathbf{C} \times \mathbf{A}.$$

Also interchanging two rows changes only the sign of the determinant, and hence

$$\mathbf{A} \cdot \mathbf{B} \times \mathbf{C} = -\mathbf{B} \cdot \mathbf{A} \times \mathbf{C}.$$

The vector triple product $\mathbf{A} \times (\mathbf{B} \times \mathbf{C})$ may be simplified by using rectangular coordinates:

$$\begin{aligned}[\mathbf{A} \times (\mathbf{B} \times \mathbf{C})]_x &= A_y(B_x C_y - B_y C_x) - A_z(B_z C_x - B_x C_z) \\ &= B_x(A_x C_x + A_y C_y + A_z C_z) - C_x(A_x B_x + A_y B_y + A_z C_z) \\ &= B_x(\mathbf{A} \cdot \mathbf{C}) - C_x(\mathbf{A} \cdot \mathbf{B}),\end{aligned}$$

and similarly for the other two components. Hence

$$\mathbf{A} \times (\mathbf{B} \times \mathbf{C}) = (\mathbf{A} \cdot \mathbf{C})\mathbf{B} - (\mathbf{A} \cdot \mathbf{B})\mathbf{C}.$$

A-3 Some identities of vector calculus

Let us first consider the divergence of the product of a scalar field w with a vector field $\mathbf{A}$. Again working in rectangular coordinates, we have

$$\begin{aligned}\operatorname{div} w\mathbf{A} &= \frac{\partial}{\partial x}(wA_x) + \frac{\partial}{\partial y}(wA_y) + \frac{\partial}{\partial z}(wA_z) \\ &= w\left(\frac{\partial A_x}{\partial x} + \frac{\partial A_y}{\partial y} + \frac{\partial A_z}{\partial z}\right) + A_x\frac{\partial w}{\partial x} + A_y\frac{\partial w}{\partial y} + A_z\frac{\partial w}{\partial z}\,; \\ \therefore \operatorname{div} w\mathbf{A} &= w \operatorname{div} \mathbf{A} + \mathbf{A}\cdot \operatorname{grad} w.\end{aligned}$$

The divergence of a cross product can be calculated in a similar manner:

$$\begin{aligned}&\operatorname{div}(\mathbf{A}\times\mathbf{B}) \\ &= \frac{\partial}{\partial x}(A_yB_z - A_zB_y) + \frac{\partial}{\partial y}(A_zB_x - A_xB_z) + \frac{\partial}{\partial z}(A_xB_y - A_yB_x) \\ &= B_x\left(\frac{\partial A_z}{\partial y} - \frac{\partial A_y}{\partial z}\right) + B_y\left(\frac{\partial A_x}{\partial z} - \frac{\partial A_z}{\partial x}\right) + B_z\left(\frac{\partial A_y}{\partial x} - \frac{\partial A_x}{\partial y}\right) \\ &\qquad - \left[A_x\left(\frac{\partial B_z}{\partial y} - \frac{\partial B_y}{\partial z}\right) + A_y\left(\frac{\partial B_x}{\partial z} - \frac{\partial B_z}{\partial x}\right) + A_z\left(\frac{\partial B_y}{\partial x} - \frac{\partial B_x}{\partial y}\right)\right]\end{aligned}$$

$$\operatorname{div}(\mathbf{A}\times\mathbf{B}) = \mathbf{B}\cdot\operatorname{curl}\mathbf{A} - \mathbf{A}\cdot\operatorname{curl}\mathbf{B}.$$

The curl operator may be treated similarly:

$$\begin{aligned}[\operatorname{curl} w\mathbf{A}]_x &= \frac{\partial}{\partial y}(wA_z) - \frac{\partial}{\partial z}(wA_y) \\ &= w\left(\frac{\partial A_z}{\partial y} - \frac{\partial A_y}{\partial z}\right) - \left(A_y\frac{\partial w}{\partial z} - A_z\frac{\partial w}{\partial y}\right) \\ &= w\,(\operatorname{curl}\mathbf{A})_x - (\mathbf{A}\times\operatorname{grad} w)_x.\end{aligned}$$

Similar relations may be obtained for the other two components. Hence

$$\operatorname{curl} w\mathbf{A} = w \operatorname{curl}\mathbf{A} - \mathbf{A}\times\operatorname{grad} w.$$

For the cross product we obtain

$$\begin{aligned}[\operatorname{curl}\mathbf{A}\times\mathbf{B}]_x &= \frac{\partial}{\partial y}(\mathbf{A}\times\mathbf{B})_z - \frac{\partial}{\partial z}(\mathbf{A}\times\mathbf{B})_y \\ &= A_x\left(\frac{\partial B_y}{\partial y} + \frac{\partial B_z}{\partial z}\right) - B_x\left(\frac{\partial A_y}{\partial y} + \frac{\partial A_z}{\partial z}\right) \\ &\quad + \left(B_y\frac{\partial}{\partial y} + B_z\frac{\partial}{\partial z}\right)A_x - \left(A_y\frac{\partial}{\partial y} + A_z\frac{\partial}{\partial z}\right)B_x.\end{aligned}$$

Adding and subtracting terms $A_x(\partial B_x/\partial x)$ and $B_x(\partial A_x/\partial x)$ and writing

$$\left(A_x\frac{\partial}{\partial x} + A_y\frac{\partial}{\partial y} + A_z\frac{\partial}{\partial z}\right) = \mathbf{A}\cdot\nabla,$$

we have

$$[\text{curl}\,\mathbf{A}\times\mathbf{B}]_x = A_x\,\text{div}\,\mathbf{B} - B_x\,\text{div}\,\mathbf{A} + (\mathbf{B}\cdot\nabla)A_x - (\mathbf{A}\cdot\nabla)B_x.$$

Similarly relations may be obtained for the other two components. Hence

$$\text{curl}\,(\mathbf{A}\times\mathbf{B}) = \mathbf{A}\,\text{div}\,\mathbf{B} - \mathbf{B}\,\text{div}\,\mathbf{A} + (\mathbf{B}\cdot\nabla)\mathbf{A} - (\mathbf{A}\cdot\nabla)\mathbf{B}.$$

Another expression used in the text, the product $\mathbf{A}\times\text{curl}\,\mathbf{A}$, may be developed as follows:

$$\begin{aligned}[\mathbf{A}\times\text{curl}\,\mathbf{A}]_x &= A_y\,\text{curl}_z\,\mathbf{A} - A_z\,\text{curl}_y\,\mathbf{A}\\ &= A_y\left(\frac{\partial A_y}{\partial x} - \frac{\partial A_x}{\partial y}\right) - A_z\left(\frac{\partial A_x}{\partial z} - \frac{\partial A_z}{\partial x}\right)\\ &= \frac{1}{2}\left(\frac{\partial A_y^2}{\partial x} + \frac{\partial A_z^2}{\partial x}\right) - \left(A_y\frac{\partial A_x}{\partial y} + A_z\frac{\partial A_x}{\partial z}\right).\end{aligned}$$

Adding a term $A_x\partial A_x/\partial x$ to each pair of parentheses yields

$$\begin{aligned}[\mathbf{A}\times\text{curl}\,\mathbf{A}]_x &= \frac{1}{2}\frac{\partial}{\partial x}(A_x^2 + A_y^2 + A_z^2) - \left(A_x\frac{\partial}{\partial x} + A_y\frac{\partial}{\partial y} + A_z\frac{\partial}{\partial z}\right)A_x\\ &= \tfrac{1}{2}\,\text{grad}_x\,A^2 - (\mathbf{A}\cdot\text{grad})\,A_x.\end{aligned}$$

Similar relations may be obtained for the other two components. Hence

$$\mathbf{A}\times\text{curl}\,\mathbf{A} = \tfrac{1}{2}\,\text{grad}\,A^2 - (\mathbf{A}\cdot\text{grad})\mathbf{A}.$$

A–4 Coordinate transformations

Cylindrical polar and spherical polar coordinates are widely used in field theory, although for the most part this text is confined to rectangular coordinates. In cylindrical polar coordinates the position of a point is defined in terms of ρ, the distance from the z-axis, the angle ϕ, and the z-coordinate as in rectangular coordinates (Fig. A-1). From the figure we can relate these quantities to the rectangular coordinates:

$$\begin{aligned}x &= \rho\cos\phi,\\ y &= \rho\sin\phi,\\ z &= z.\end{aligned}$$

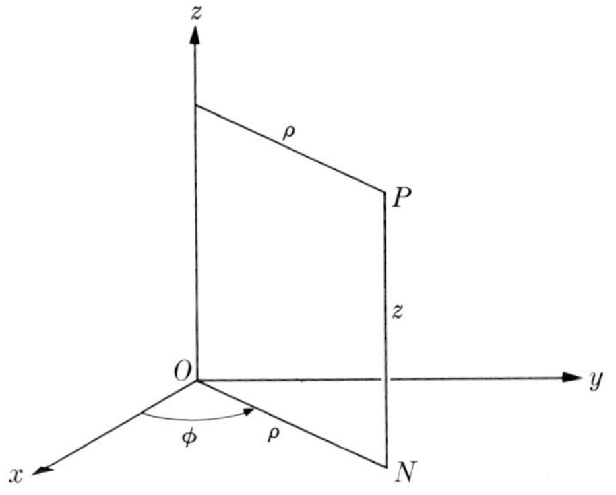

FIG. A–1. The transformation from rectangular to cylindrical polar coordinates.

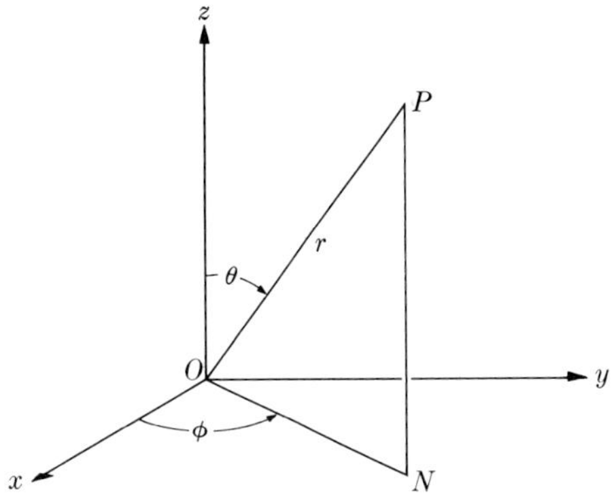

FIG. A–2. The transformation from rectangular to spherical polar coordinates.

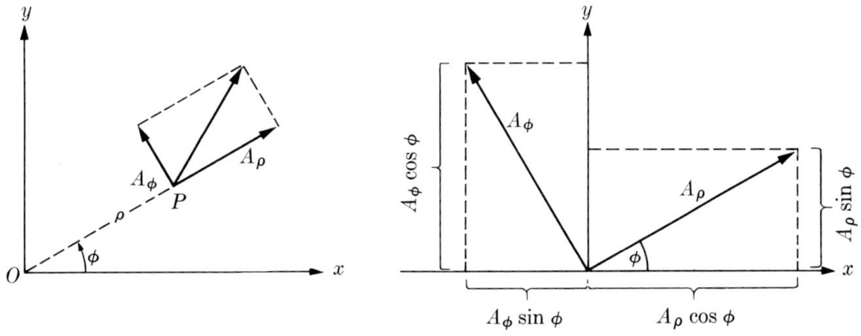

FIG. A–3. The relation between the rectangular and cylindrical components of a vector.

In spherical polars the coordinates are (r, θ, ϕ) as shown in Fig. A-2, and we have

$$x = r \sin \theta \cos \phi,$$

$$y = r \sin \theta \sin \phi,$$

$$z = r \cos \theta.$$

It is also important to be able to express the components of a vector in one coordinate system in terms of the components in another. Fig. A-3 shows a sketch of the xy-plane in cylindrical coordinates, with the vector at the point P resolved into components in the cylindrical system. From the figure we see that

$$A_x = A_\rho \cos \phi - A_\phi \sin \phi,$$

$$A_y = A_\rho \sin \phi + A_\phi \cos \phi,$$

and the z-component is the same in both systems. Similarly, for spherical polar coordinates, we have

$$A_x = A_r \sin\theta \cos\phi + A_\theta \cos\theta \cos\phi - A_\phi \sin\phi,$$

$$A_y = A_r \sin\theta \sin\phi + A_\theta \cos\theta \sin\phi + A_\phi \cos\phi,$$

$$A_z = A_r \cos\theta - A_\theta \sin\theta.$$

The inverse relations are also of interest:

$$A_r = A_x \sin\theta \cos\phi + A_y \sin\theta \sin\phi + A_z \cos\theta,$$

$$A_\theta = A_x \cos\theta \cos\phi + A_y \cos\theta \sin\phi - A_z \sin\theta,$$

$$A_\phi = -A_x \sin\phi + A_y \cos\phi.$$

A-5 Vector operations in polar coordinates

We defined the operations grad, div, and curl without reference to any coordinate system. We can express them in terms of the polar systems, using the original definitions and methods analogous to those used in expressing these operations in rectangular coordinates. Let us consider, for example, the expression of div in cylindrical coordinates. The volume element is made up as shown in Fig. A-4. We first consider the flow over the two faces perpendicular to $\mathbf{u}_\rho$. Over the inner face this will be $-A_\rho \rho \, d\phi \, dz$ and the outer face $[A_\rho + (\partial A_\rho/\partial\rho)\, d\rho](\rho + d\rho)\, d\phi \, dz$. Adding and retaining small quantities up to third order, we find that the net flow becomes

$$A_\rho \, d\rho \, d\phi \, dz + \frac{\partial A_\rho}{\partial \rho} \rho \, d\rho \, d\phi \, dz.$$

A similar expression may be obtained for the other two pairs of faces. The total net outflow may be written as

$$A_\rho \, d\rho \, d\phi \, dz + \frac{\partial A_\rho}{\partial \rho} \rho \, d\rho \, d\phi \, dz + \frac{\partial A_\phi}{\partial \phi} d\rho \, d\phi \, dz + \frac{\partial A_z}{\partial z} \rho \, d\rho \, d\phi \, dz.$$

The volume of the element is given by $\rho \, d\rho \, d\phi \, dz$, and hence

$$\operatorname{div} \mathbf{A} = \frac{A_\rho}{\rho} + \frac{\partial A_\rho}{\partial \rho} + \frac{1}{\rho}\frac{\partial A_\phi}{\partial \phi} + \frac{\partial A_z}{\partial z}$$

or

$$\operatorname{div} \mathbf{A} = \frac{1}{\rho}\frac{\partial}{\partial \rho}(\rho A_\rho) + \frac{1}{\rho}\frac{\partial A_\phi}{\partial \phi} + \frac{\partial A_z}{\partial z}.$$

The quantity curl **A** can also be obtained by a precisely analogous argument.

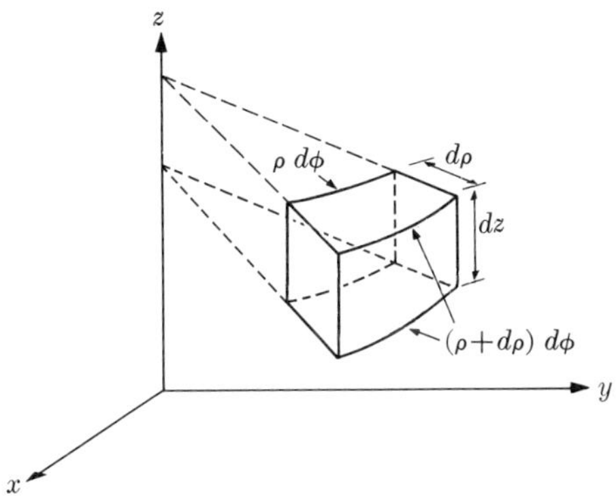

FIG. A–4. The volume element in cylindrical polar coordinates.

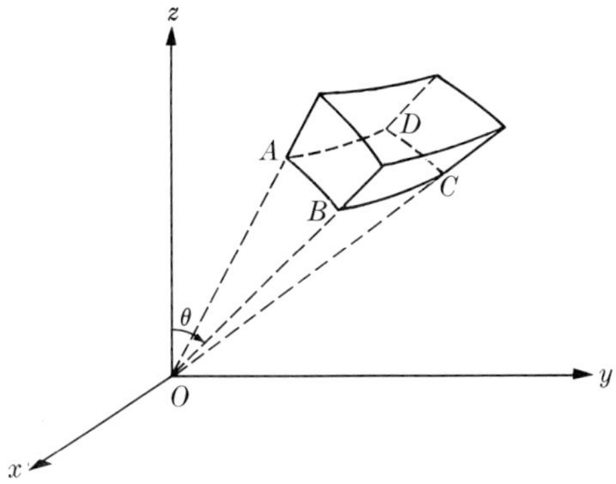

FIG. A–5. The volume element in spherical polar coordinates.

The result is

$$\operatorname{curl} \mathbf{A} = \mathbf{u}_\rho \left(\frac{1}{\rho} \frac{\partial A_z}{\partial \phi} - \frac{\partial A_\phi}{\partial z} \right) + \mathbf{u}_\phi \left(\frac{\partial A_\rho}{\partial z} - \frac{\partial A_z}{\partial \rho} \right) + \mathbf{u}_z \left[\frac{1}{\rho} \frac{\partial}{\partial \rho} (\rho A_\phi) - \frac{1}{\rho} \frac{\partial A_\rho}{\partial \phi} \right].$$

The volume element in spherical polar coordinates is shown in Fig. A-5. In this case, for divergence we have

$$\operatorname{div} \mathbf{A} = \frac{1}{r^2} \frac{\partial}{\partial r} (r^2 A_r) + \frac{1}{r \sin\theta} \frac{\partial}{\partial \theta} (A_\theta \sin\theta) + \frac{1}{r \sin\theta} \frac{\partial A_\phi}{\partial \phi}.$$

We take the calculation of the r-component of curl as our next example. We must first calculate the line integral around the curvilinear contour $ABCD$. Note that this is perpendicular to $\mathbf{u}_r$, and the sense is related to it by the right-hand rule. For the lengths of the sides, we have

$$AB = r\, d\theta, \qquad CD = r\, d\theta, \qquad BC = r \sin(\theta + d\theta)\, d\phi, \qquad DA = r \sin\theta\, d\phi.$$

The line integral is then given by

$$A_\theta r\, d\theta + \left(A_\phi + \frac{\partial A_\phi}{\partial \theta} d\theta \right) r \sin(\theta + d\theta)\, d\phi - \left(A_\theta + \frac{\partial A_\theta}{\partial \phi} d\phi \right) r\, d\theta$$
$$- A_\phi r \sin\theta\, d\phi = - \frac{\partial A_\theta}{\partial \phi} r\, d\theta\, d\phi + \frac{\partial}{\partial \theta} (A_\phi \sin\theta) r\, d\theta\, d\phi,$$

where small quantities have been retained up to second order. The area $ABCD$ is given by $r^2 \sin\theta\, d\theta\, d\phi$, and hence

$$\operatorname{curl}_r \mathbf{A} = \frac{1}{r \sin\theta} \left[\frac{\partial}{d\theta} (A_\phi \sin\theta) - \frac{\partial A_\theta}{\partial \phi} \right].$$

The other two components may be obtained in a similar manner, and the complete result is

$$\operatorname{curl}\mathbf{A} = \mathbf{u}_r \frac{1}{r\sin\theta}\left[\frac{\partial}{\partial\theta}(A_\phi \sin\theta) - \frac{\partial A_\theta}{\partial\phi}\right] + \mathbf{u}_\theta \frac{1}{r}\left[\frac{1}{\sin\theta}\frac{\partial A_r}{\partial\phi} - \frac{\partial}{\partial r}(rA_\phi)\right] + \mathbf{u}_\phi \frac{1}{r}\left[\frac{\partial}{\partial r}(rA_\theta) - \frac{\partial A_r}{\partial\theta}\right].$$

Finally we express the operation ∇^2 in spherical polar coordinates. The components of gradient are given by the change in the variable V divided by the element of length in the appropriate direction. Thus

$$\operatorname{grad} V = \mathbf{u}_r \frac{\partial V}{\partial r} + \mathbf{u}_\theta \frac{1}{r}\frac{\partial V}{\partial\theta} + \mathbf{u}_\phi \frac{1}{r\sin\theta}\frac{\partial V}{\partial\phi}.$$

Substituting these components into the expression given above for div $\mathbf{A}$, we readily obtain

$$\operatorname{div}\operatorname{grad} V = \nabla^2 V = \frac{1}{r^2}\frac{\partial}{\partial r}\left(r^2\frac{\partial V}{\partial r}\right) + \frac{1}{r^2\sin\theta}\frac{\partial}{\partial\theta}\left(\sin\theta\frac{\partial V}{\partial\theta}\right) + \frac{1}{r^2\sin^2\theta}\frac{\partial^2 V}{\partial\phi^2}.$$

Index

ABCDE698765